WRITING TESTBENCHES

Functional Verification of HDL Models

WRITING TESTBENCHES

Functional Verification of HDL Models

Janick Bergeron
Qualis Design Corporation

KLUWER ACADEMIC PUBLISHERS
Boston / Dordrecht / London

Distributors for North, Central and South America:
Kluwer Academic Publishers
101 Philip Drive, Assinippi Park
Norwell, Massachusetts 02061 USA
Telephone (781) 871-6600
Fax (781) 871-6528
E-Mail <kluwer@wkap.com>

Distributors for all other countries:
Kluwer Academic Publishers Group
Distribution Centre
Post Office Box 322
3300 AH Dordrecht, THE NETHERLANDS
Telephone 31 78 6392 392
Fax 31 78 6546 474
E-Mail <orderdept@wkap.nl>

Electronic Services <http://www.wkap.nl>

Library of Congress Cataloging-in-Publication Data
Bergeron, Janick.
    Writing testbenches : functional verification of HDL models / Janick Bergeron.
        p. cm.
    ISBN 0-7923-7766-4 (alk. paper)
    1. Computer hardware description languages. 2. Integrated circuits--Verification. I.
Title.

    TK7885.7.B47 2000
    621.3815--dc21   99-089335

TABLE OF CONTENTS

About the Cover   *xiii*

Foreword   *xv*

Preface   *xvii*

Why This Book Is Important xvii
What This Book Is About xviii
What Prior Knowledge You Should Have xix
Reading Paths xx
VHDL versus Verilog xx
For More Information xxii
Acknowledgements xxii

CHAPTER 1   *What is Verification?*   *1*

What is a Testbench? 1
The Importance of Verification 2
Reconvergence Model 4
The Human Factor 5
  Automation *6*
  Poka-Yoka *6*

Redundancy *6*

What Is Being Verified? 7

Formal Verification *7*

Equivalence Checking *8*

Model Checking *9*

Functional Verification *10*

Testbench Generation *11*

Functional Verification Approaches 11

Black-Box Verification *12*

White-Box Verification *13*

Grey-Box Verification *13*

Testing Versus Verification 13

Scan-Based Testing *14*

Design for Verification *16*

Verification and Design Reuse 16

Reuse Is About Trust *16*

Verification for Reuse *17*

The Cost of Verification 17

Summary 19

CHAPTER 2   *Verification Tools*   *21*

Linting Tools 22

The Limitations of Linting Tools *23*

Linting Verilog Source Code *25*

Linting VHDL Source Code *26*

Code Reviews *28*

Simulators 28

Stimulus and Response *29*

Event-Driven Simulation *29*

Cycle-Based Simulation *31*

Co-Simulators *34*

Third-Party Models 36

Hardware Modelers *37*

Waveform Viewers 38

Code Coverage 40

Statement Coverage *42*

Path Coverage *44*

Expression Coverage *45*

What Does 100 Percent Coverage Mean? 45

Verification Languages 46

Revision Control 47

The Software Engineering Experience 48

Configuration Management 50

Working with Releases 51

Issue Tracking 52

What Is an Issue? 52

The Grapevine System 53

The Post-It System 54

The Procedural System 55

Computerized System 55

Metrics 57

Code-Related Metrics 57

Quality-Related Metrics 58

Interpreting Metrics 59

Summary 60

CHAPTER 3   *The Verification Plan*   *61*

The Role of the Verification Plan 62

Specifying the Verification 62

Defining First-Time Success 63

Levels of Verification 64

Unit-Level Verification 65

Reusable Components Verification 66

ASIC and FPGA Verification 67

System-Level Verification 68

Board-Level Verification 68

Verification Strategies 69

Verifying the Response 70

Random Verification 71

From Specification to Features 72

Component-Level Features 73

System-Level Features 73

Error Types to Look For 74

From Features to Testcases 75

Prioritize 75

Group into Testcases 76

Design for Verification 77

From Testcases to Testbenches **79**

Verifying Testbenches *80*

Summary **81**

CHAPTER 4 *Behavioral Hardware Description*
Languages *83*

Behavioral versus RTL Thinking **83**

Contrasting the Approaches *85*

You Gotta Have Style! **87**

A Question of Discipline *87*

Optimize the Right Thing *88*

Good Comments Improve Maintainability *91*

Structure of Behavioral Code **92**

Encapsulation Hides Implementation Details *93*

Encapsulating Useful Subprograms *94*

Encapsulating Bus-Functional Models *97*

Data Abstraction **100**

Real Values *101*

Records *105*

Multi-Dimensional Arrays *112*

Lists *115*

Files *121*

Interfacing High-Level Data Types *124*

The HDL Parallel Engine **125**

Connectivity, Time, and Concurrency.. .... *125*

Connectivity, Time, and Concurrency in HDLs *126*

The Problems with Concurrency *127*

Emulating Parallelism on a Sequential Processor ... *128*

The Simulation Cycle *129*

Parallel vs. Sequential *132*

Fork/Join Statement *134*

The Difference Between Driving and Assigning *137*

Verilog Portability Issues **140**

Read/Write Race Conditions *141*

Write/Write Race Conditions *144*

Initialization Races *146*

Guidelines for Avoiding Race Conditions *147*

Events from Overwritten Scheduled Values *147*

Disabled Scheduled Values *148*

Output Arguments on Disabled Tasks *150*

Non-Reentrant Tasks *151*

Summary **153**

CHAPTER 5 *Stimulus and Response* **155**

Simple Stimulus **155**

Generating a Simple Waveform *156*

Generating a Complex Waveform *159*

Generating Synchronized Waveforms *160*

Aligning Waveforms in Delta-Time *164*

Generating Synchronous Data Waveforms *165*

Encapsulating Waveform Generation *167*

Abstracting Waveform Generation *169*

Verifying the Output **172**

Visual Inspection of Response *172*

Producing Simulation Results *172*

Minimizing Sampling *174*

Visual Inspection of Waveforms *174*

Self-Checking Testbenches **176**

Input and Output Vectors *176*

Golden Vectors *177*

Run-Time Result Verification *179*

Complex Stimulus **183**

Feedback Between Stimulus and Design *183*

Recovering from Deadlocks *184*

Asynchronous Interfaces *187*

CPU Operations *189*

Configurable Operations *192*

Complex Response **193**

What is a Complex Response? *194*

Handling Unknown or Variable Latency *195*

Abstracting Output Operations *199*

Generic Output Monitors *202*

Monitoring Multiple Possible Operations *203*

Monitoring Bi-Directional Interfaces *205*

Predicting the Output **211**

Data Formatters *211*

Packet Processors *215*

Complex Transformations *216*

Summary **219**

CHAPTER 6 *Architecting Testbenches* *221*

Reusable Verification Components **221**
  Procedural Interface *225*
  Development Process *226*
Verilog Implementation **227**
  Packaging Bus-Functional Models *228*
  Utility Packages *231*
VHDL Implementation **237**
  Packaging Bus-Functional Procedures *238*
  Creating a Test Harness *240*
  Abstracting the Client/Server Protocol *243*
  Managing Control Signals *246*
  Multiple Server Instances *247*
  Utility Packages *249*
Autonomous Generation and Monitoring **250**
  Autonomous Stimulus *250*
  Random Stimulus *253*
  Injecting Errors *255*
  Autonomous Monitoring *255*
  Autonomous Error Detection *258*
Input and Output Paths **258**
  Programmable Testbenches *259*
  Configuration Files *260*
  Concurrent Simulations *261*
  Compile-Time Configuration *262*
Verifying Configurable Designs **263**
  Configurable Testbenches *265*
  Top Level Generics and Parameters *266*
Summary **268**

CHAPTER 7 *Simulation Management* *269*

Behavioral Models **269**
  Behavioral versus Synthesizable Models *270*
  Example of Behavioral Modeling *271*
  Characteristics of a Behavioral Model *273*

Modeling Reset 276
Writing Good Behavioral Models 281
Behavioral Models Are Faster 285
The Cost of Behavioral Models 286
The Benefits of Behavioral Models 286
Demonstrating Equivalence 289

Pass or Fail? 289
Managing Simulations 292
Configuration Management 294
Verilog Configuration Management 295
VHDL Configuration Management 301
SDF Back-Annotation 305
Output File Management 309

Regression 312
Running Regressions 313
Regression Management 314

Summary 316

APPENDIX A *Coding Guidelines* *317*

Directory Structure 318
VHDL Specific 320
Verilog Specific 320

General Coding Guidelines 321
Comments 321
Layout 323
Syntax 326
Debugging 329

Naming Guidelines 329
Capitalization 330
Identifiers 332
Constants 334
HDL Specific 334
Filenames 336

HDL Coding Guidelines 336
Structure 337
Layout 337
VHDL Specific 337
Verilog Specific 340

Afterwords   ***347***

Index   ***349***

ABOUT THE COVER

The Quebec Bridge Company was formed in 1887 and for the next thirteen years, very little was done. In 1900, Thomas Cooper, a consultant in civil engineering specializing in bridge building was appointed the company's consulting engineer for the duration of the work. The Sixth Street bridge in Pittsburgh, the Seekonk bridge in Providence, and the Second Avenue bridge in New York were already part of his portfolio.

Conscious of the precarious financial situation of the Quebec Bridge Company, Cooper recommended the cantilever superstructure proposed by the Phoenix Bridge Company, of Phoenixville, Pennsylvania as the best and cheapest of the proposals. He also recommended that the span of the bridge be increased from 1600 feet to 1800 feet to minimize the cost of constructing the piers supporting the bridge. The specifications were also modified to allow for greater unit stresses in the structure. The Quebec Bridge was to be the longest cantilever bridge in the world.

For the next three years, the assumptions underlying the modified design of the most technically ambitious bridge in the world went unchallenged. After the Canadian government guaranteed a bond issue in 1903, the construction shifted into high gear. In the rush to minimize delays, the assumed weight of the revised bridge was not recalculated. Instead, work continued with the estimated weight the Phoenix Company had provided with the original proposal. Cooper was personally offended when the Canadian Department of Railways and Canals requested that the plans be independently reviewed and approved. With full confidence in his own design and expertise, Cooper managed to turn the independent review into a rubber stamp approval by unqualified individuals.

Subsequent warnings were also summarily ignored. In 1906, the Phoenix Company's inspector of material reported that the actual weight of steel put into the bridge had already exceeded the original estimated weight. The final weight of the bridge was estimated to be eleven million pounds higher than originally thought. The alternative being to start building the bridge all over again, Cooper concluded that the increase in stresses was acceptable.

In early August 1907, the lower horizontal pieces running the length of the bridge began to show signs of buckling. The Phoenix Company insisted that they were already bent when they left the shop in Phoenixville and work continued. They made no effort to explain why a deflection had increased by an inch and a half in the past week. On August 29th, the south arm of the bridge collapsed under its own weight, killing 73 workers.

The bridge was eventually redesigned and rebuilt, weighing two and a half times more than its predecessor. Ironically, it suffered a tragedy of its own in 1916 when the central span fell into the river while it was being hoisted into place. The bridge was finally completed in 1918[1]. It is still in use today and it is still the longest cantilever bridge in the world.

The parallel with today's micro-electronic designs is obvious. The next design is always more challenging than the previous one and it takes the designers into previously uncharted waters. A design cannot go unchallenged simply because it worked in the last implementation. Changes in the specifications or new functionality cannot be assumed to work simply because they were designed by the best engineers. Errors will be made. It is important that any design be independently verified to ensure that it is indeed functionally correct. The alternative is to manufacture a non-functional design. Hopefully, no one will get killed. But many could lose their jobs.

---

1. The technical drawing on the cover, from the St. Lawrence Bridge Company, is of that newer bridge, copyright ADF Industries Lourdes, Lachine, Qc, Canada. All Rights Reserved. Photo: CP Picture Archive. Cover designed by Elizabeth Nephew (www.nephco.com).

FOREWORD

With gate counts and system complexity growing exponentially, engineers confront the most perplexing challenge in product design: functional verification. The bulk of the time consumed in the design of new ICs and systems is now spent on verification. New and interesting design technologies like physical synthesis and design reuse that create ever-larger designs only aggravate the problem. What the EDA tool industry has continuously failed to realize is that the real problem is not how to *create* a 12 million gate IC that runs at 600 MHz, but how to *verify* it.

The true path to rapid and accurate system verification includes both tool and methodology innovation. Engineers are compelled to use the best verification and design tools available to shorten design cycle time. But it is a grave mistake to assume that simply having the best tools will result in quality ICs and systems that meet market demands and mitigate competitive pressures. Indeed, the best determinant of successful IC and system design is how the engineer approaches the problem. The vast majority of engineers still approach verification using methods dating back a dozen years or more. If dramatic improvement in verification efficiency is to come about, real change in the way engineers approach verification is mandatory and urgent.

A review of available texts on verification reveals a lack of mature knowledge and leadership on the topic: the EDA tool industry is conspicuously silent. This text marks the first genuine effort at defining a verification methodology that is independent of both tools and applications. Engineers now have a true reference text for quickly and accurately verifying the functionality of their designs. Experts may disagree on the specifics of implementing the methodology, but the fact still remains that a mature verification methodology like this is long overdue.

A reasonable question to ask is, why develop a verification methodology that is tool-independent? Simply put, tool-independent methodologies yield more predictable and adaptable processes, have lower adoption costs, and are not dependent upon the continued technical superiority of any one EDA tool vendor. As has been proven so many times in the EDA tool industry, today's technology leader can quickly become tomorrow's lost cause. An independent methodology yields the ability to tap the best of today's EDA tools, while retaining the powerful option of switching to competing products as they become technically or financially superior.

The future of system verification undoubtedly will include new and innovative approaches and more powerful tools. Much work remains in the area of hardware/software co-verification. New verification-centric languages that leverage object-oriented techniques hold great promise. However, the best verification methodologies of the future will come from a tool-independent view, where the mind is free to dream up new and innovative ways to tackle verification. This book is a necessary and great beginning.

Michael Horne, President & CEO
Qualis Design Corporation

PREFACE

If you survey hardware design groups, you will learn that between 60 and 80 percent of their effort is now dedicated to verification. Unlike synthesizeable coding, there is no particular coding style required for verification. The freedom of using any features of the languages has produced a wide array of techniques and approaches to verification. The absence of constraints and lack of available expertise and references in verification has resulted in *ad hoc* approaches. The consequences of an informal verification process can range from a non-functional design requiring several re-spins, through a design with only a subset of the intended functionality, to a delayed product shipment.

WHY THIS BOOK IS IMPORTANT

Take a survey of the books about Verilog or VHDL currently available. You will notice that the majority of the pages are devoted to explaining the details of the languages. In addition, several chapters are focused on the synthesizeable coding style - or RTL - replete with examples. Some books are even devoted entirely to the subject of RTL coding.

When verification is addressed, only one or two chapters are dedicated to the topic. And often, the primary focus is to introduce more language constructs. Verification is often presented in a very rudimentary fashion, using simple techniques that become tedious in large-scale, real-life designs.

Basic language textbooks appeared early in the life of hardware description languages. They continue to appear as the presentation styles are refined to facilitate their understanding or as they are tuned to a particular audience. But the benefits of additional language textbooks is quickly diminishing.

Since the introduction of hardware description languages and logic synthesis technology in the mid 80's, a lot of expertise on coding styles and synthesis methodologies has been developed to obtain desired results. A lot of that expertise has been presented at conferences, has been codified in textbooks, and is available as introductory and advanced training classes. Early language-focused textbooks were re-edited to include chapters on synthesizeable coding and many books entirely devoted to that topic now exist. The synthesis process, although complex, is now well understood and many resources are available to communicate that understanding. Standard procedures and techniques have been developed to predictably produce adequate results.

At the time I started writing this book, it was going to be the first book specifically devoted to verification techniques for hardware models. I will introduce you to the latest verification techniques that have been successfully used to produce first-time-right ASICs, Systems-on-a-Chip (SoC), boards, and entire systems.

WHAT THIS BOOK IS ABOUT

I will first introduce the necessary concepts and tools of verification, then I'll describe a process for carrying out an effective functional verification of a design. It is necessary to cover some language semantics that are often overlooked or oversimplified in textbooks intent on describing the synthesizeable subset. These unfamiliar semantics become important in understanding what makes a well-implemented and robust testbench and in providing the necessary control and monitor features.

I will also present techniques for applying stimulus and monitoring the response of a design, by abstracting the operations using bus-functional models. The architecture of testbenches built around these bus-functional models is important to minimize development and maintenance effort.

Behavioral modeling is another important concept presented in this book. It is used to parallelize the implementation and verification of a design and to perform more efficient simulations. For many, behavioral modeling is synonymous with synthesizeable or RTL modeling. In this book, the term "behavioral" is used to describe any model that adequately emulates the functionality of a design, usually using non-synthesizeable constructs and coding style.

WHAT PRIOR KNOWLEDGE YOU SHOULD HAVE

This book focuses on the functional verification of hardware designs using *either* VHDL or Verilog. I expect the reader to have at least a basic knowledge of one of the languages. Ideally, you should have experience in writing synthesizeable models and be familiar with running a simulation using any of the available VHDL or Verilog simulators. There will be no description of language syntax or grammar. It may be a good idea to have a copy of a language-focused textbook as a reference along with this book[1]. I do not describe the synthesizeable subset, nor limit the implementation of the verification techniques to using that subset. Verification is a complex task: the power of either language will be used to their fullest.

I also expect that you have a basic understanding of digital hardware design. This book uses several hypothetical designs from various domains of application (video, datacom, computing, etc.). How these designs are actually specified, architected, then implemented is beyond the scope of this book. The content focuses on the specification, architecture, then implementation of the *verification* of these same designs.

---

1. For Verilog, I recommend *The Verilog Hardware Description Language* by Thomas & Moorby, 3rd edition or later (Kluwer Academic Publisher). For VHDL, I recommend *VHDL Coding Styles and Methodologies* by Ben Cohen (Kluwer Academic Publisher).

READING PATHS

You should really read this book from cover to cover. However, if you are pressed for time, here are a few suggested paths.

If you are using this book as a university or college textbook, you should focus on Chapter 4, Chapter 5, and Appendix A. If you are a junior engineer who has only recently joined a hardware design group, you may skip Chapters 3, 6 and 7. But do not forget to read them once you have gained some experience.

Chapters 3 and 6, as well as Appendix A, will be of interest to a senior engineer in charge of defining the verification strategy for a project. If you are an experienced designer, you may wish to skip ahead to Chapter 3. If you are an experienced Verilog or VHDL user, you may wish to skip Chapter 4 - but read it anyway, just to make sure your definition of "*experienced*" matches mine.

If you have a software background, Chapter 7 and Appendix A may seem somewhat obvious. If you have a hardware design and RTL coding mindset, Chapters 4 and 7 are probably your best friends.

If your responsibilities are limited to managing a hardware verification project, you probably want to concentrate on Chapter 3, Chapter 6, and Chapter 7.

VHDL VERSUS VERILOG

The first decision a design group is often faced with is deciding which language to use. As the author of this book, I faced the same dilemma. The answer is usually dictated by the individual's own knowledge or personal preference.

I know both languages equally well. I work using both. I teach them both. When asked which one I prefer, I usually answer that I was asked the wrong question. The right question should be "*Which one do I hate the least?*" And the answer to that question is: "*the one I'm not currently working with*".

When working in one language, you do not notice the things that are simple to describe or achieve in that language. Instead, you notice the frustrations and how it would be easy to do it if only you were using the *other* language.

In my opinion, both languages are inadequate by themselves, especially for verification. They are both equally poor for synthesizeable description. Some things are easier to accomplish in one language than in the other. For a *specific* model, one language is better than the other: one language has features that better map to the functionality to be modeled. However, as a general rule, neither is better than the other.

Verification techniques transcend the language used. VHDL and Verilog are only implementation vehicles. Both are used throughout the book. It is not to say that this book is bilingual: examples are shown in only one language. I trust that a VHDL-only or Verilog-only reader will be able to understand the example in the other language, even though the syntax is slightly different.

Some sections are Verilog only. In my experience Verilog is a much abused language. It has the reputation for being easier to learn than VHDL, and to the extent that the learning curve is not as steep, it is true. However, both languages provide similar concepts: sequential statements, parallel constructs, structural constructs, and the illusion of parallelism.

For both languages, these concepts must be learned. Because of its lax requirements, Verilog lulls the user into a false sense of security. The user believes that he or she knows the language because there are no syntax errors or because the simulation results appear to be correct. Over time, and as a design grows, race conditions and fragile code structures become apparent, forcing the user to learn these important concepts. Both languages have the same *area under the learning curve*. VHDL's is steeper but Verilog's goes on for much longer. Some sections in this book take the reader further down the Verilog learning curve.

FOR MORE INFORMATION

If you want more information on topics mentioned in this book, you will find links to relevant resources at the following URL:

`http://janick.bergeron.com/wtb`

In the *resources* area, you will find links to publicly available utilities, documents and tools that make the verification task easier. You will also find an errata section listing and correcting the errors that inadvertently made their way in the book.[2]

The website also has a set of quizzes corresponding to each chapter in this book. I recommend you complete them to verify your understanding of the material. You may even choose to start with a quiz to determine if you need to read a chapter or not!

ACKNOWLEDGEMENTS

My wife, Danielle, gave this book energy against its constant drain. Kyle Smith, my editor, gave it form from its initial chaos. Ben Cohen, Ken Coffman, and Bernard Delay, my technical reviewers, gave it light from its initial darkness. And FrameMaker, my word processing software, reliably went where no Word had gone before!

---

2. If you know of a verification-related resource or an error in this book that is not mentionned in the website, please let me know via email at `janick@bergeron.com`. I thank you in advance.

CHAPTER 1   WHAT IS VERIFICATION?

Verification is not a testbench, nor is it a series of testbenches. Verification is a *process* used to demonstrate the functional correctness of a design. We all perform verification processes throughout our daily lives: balancing a checkbook, tasting a simmering dish, associating landmarks with symbols on a map. These are all verification processes.

In this chapter, I introduce the basic concepts of verification, from its importance and cost, to making sure you are verifying that you are indeed implementing what you want. We look at the differences between various verification approaches as well as the difference between testing and verification. I also show how verification is key to design reuse.

WHAT IS A TESTBENCH?

The term "testbench", in VHDL and Verilog, usually refers to the code used to create a pre-determined input sequence to a design, then optionally observe the response. It is commonly implemented using VHDL or Verilog, but may also include external data files or C routines.

Figure 1-1 shows how a testbench interacts with a *Design Under Verification* (DUV). The testbench provides inputs to the design and monitors any outputs. Notice how this is a completely closed system: no inputs or outputs go in or out. The testbench is effectively a model of the universe as far as the design is concerned. The verification challenge is to determine what input patterns to supply to the design and what is the expected output of a properly working design.

Figure 1-1. Generic structure of a testbench and design under test

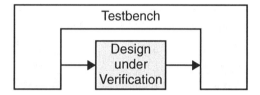

THE IMPORTANCE OF VERIFICATION

Most books focus on syntax, semantic and RTL subset.

If you look at a typical book on Verilog or VHDL, you will find that most of the chapters are devoted to describing the syntax and semantics of the language. You will also invariably find two or three chapters on synthesizeable coding style or Register Transfer Level (RTL) subset.

Most often, only a single chapter is dedicated to testbenches. Very little can be adequately explained in one chapter and these explanations are usually very simplistic. In nearly all cases, these books limit the techniques described to applying simple sequences of vectors in a synchronous fashion. The output is then verified using a waveform viewing tool. Most also take advantage of the topic to introduce the file input mechanisms offered by the language, devoting yet more content to detailed syntax and semantics.

Given the significant proportion of literature devoted to writing synthesizeable VHDL or Verilog code compared to writing testbenches to verify their functional correctness, you could be tempted to conclude that the former is a more daunting task than the latter. The evidence found in all hardware design teams points to the contrary.

70% of design effort goes to verification.

Today, in the era of multi-million gate ASICs, reusable Intellectual Property (IP), and System-on-a-Chip (SoC) designs, verification consumes about 70% of the design effort. Design teams, properly staffed to address the verification challenge, include engineers dedicated to verification. The number of verification engineers is usually twice the number of RTL designers. When design projects are completed, the code that implements the testbenches makes up to 80% of the total code volume.

Verification is on critical path.

Given the amount of effort demanded by verification, the shortage of qualified hardware design and verification engineers, and the quantity of code that must be produced, it is no surprise that, in all projects, verification rests squarely on the critical path. It is also the reason verification is currently the target of new tools and methodologies. These tools and methodologies attempt to reduce the overall verification time by enabling parallelism of effort, higher levels of abstraction and automation.

Verification time can be reduced through parallelism.

If effort can be parallelized, additional resources can be applied effectively to reduce the total verification time. For example, digging a hole in the ground can be parallelized by providing more workers armed with shovels. To parallelize the verification effort, it is necessary to be able to write - and debug - testbenches in parallel with each others as well as in parallel with the implementation of the design.

Verification time can be reduced through abstraction.

Providing higher levels of abstraction enables you to work more efficiently without worrying about low-level details. Using a backhoe to dig the same hole mentioned above is an example of using a higher level of abstraction.

Using abstraction reduces control over low-level details.

Higher levels of abstraction are usually accompanied by a reduction in control and therefore must be chosen wisely. These higher levels of abstraction often require additional training to understand the abstraction mechanism and how the desired effect can be produced.

Using a backhoe to dig a hole suffers from the same loss-of-control problem: the worker is no longer directly interacting with the dirt; instead the worker is manipulating levers and pedals. Digging happens much faster, but with lower precision and only by a trained operator. The verification process can use higher levels of abstraction by working at the transaction- or bus-cycle-levels (or even higher ones), instead of always dealing with low-level zeroes and ones.

Verification time can be reduced through automation.

Automation lets you do something else while a machine completes a task autonomously, faster, and with predictable results. Automation requires standard processes with well-defined inputs and outputs. Not all processes can be automated. Holes must be dug in a variety of shapes, sizes, depths, locations, and in varying soil conditions, which render general-purpose automation impossible.

Verification faces similar challenges. Because of the variety of functions, interfaces, protocols, and transformations that must be verified, it is not possible to provide a general purpose automation solution for verification given today's technology. It is possible to automate some portion of the verification process, especially when applied to a narrow application domain. For example, trenchers have automated digging holes used to lay down conduits or cables at shallow depths. Tools automating various portions of the verification process will be introduced. Hopefully, this book may help define new standard verification practices that could be automated in a near future.

RECONVERGENCE MODEL

The *reconvergence model* is a conceptual representation of the verification process. It is used to illustrate what exactly is being verified.

Do you know what you are actually verifying?

One of the most important questions you must be able to answer is: *What are you verifying*? The purpose of verification is to ensure that the result of some transformation is as intended or as expected. For example, the purpose of balancing a checkbook is to ensure that all transactions have been recorded accurately and confirm that the balance in the register reflects the amount of available funds.

Figure 1-2.
Reconvergent paths in verification

Verification is the reconciliation, through different means, of a specification and an output.

Figure 1-2 shows that verification of a transformation can only be accomplished through a second reconvergent path with a common source. The transformation can be any process that takes an input and produces an output. RTL coding from a specification, insertion of a scan chain, synthesizing RTL code into a gate-level netlist, and layout of a gate-level netlist are some of the transformations performed in a hardware design project. The verification process reconciles the result with the starting point. If there is no starting point common to the transformation and the verification, <u>no</u> verification takes place.

The reconvergent model can be described using the checkbook example as is illustrated in Figure 1-3. The common origin is the previous month's balance in the checking account. The transformation is the writing, recording and debiting of several checks during a one-month period. The verification reconciles the final balance in the checkbook register using this month's bank statement.

Figure 1-3.
Balancing a
checkbook is a
verification
process

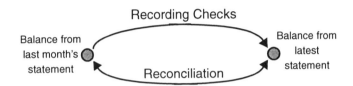

THE HUMAN FACTOR

If the transformation process is not completely automated from end to end, it is necessary for an individual (or group of individuals) to interpret a specification of the desired outcome then perform the transformation. RTL coding is an example of this situation. A design team interprets a written specification document and produces what they believe to be functionally correct synthesizeable HDL code. Usually, each engineer is left to verify that the code written is indeed functionally correct.

Figure 1-4.
Reconvergent
paths in
ambiguous
situation

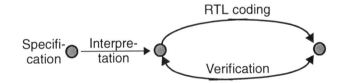

A designer verifying his or her own design verifies against his or her own interpretation, not against the specification.

Figure 1-4 shows the reconvergent path model of the situation described above. If the same individual performs the verification of the RTL coding that initially required interpretation of a specification, then the common origin is that interpretation, <u>not</u> the specification.

In this situation, the verification effort verifies whether the design accurately represents the *implementer's interpretation* of that specification. If that interpretation is wrong in any way, then this verification activity will never highlight it.

Any human intervention in a process is a source of uncertainty and unrepeatability. The probability of human-introduced errors in a process can be reduced through several complementary mechanisms: automation, *poka-yoka*, or redundancy.

Automation

Eliminate human intervention.

Automation is the obvious way to eliminate human-introduced errors in a process. Automation takes human intervention completely out of the process. However, automation is not always possible, especially in processes that are not well-defined and continue to require human ingenuity and creativity, such as hardware design.

Poka-Yoka

Make human intervention foolproof.

Another possibility is to mistake-proof the human intervention by reducing it to simple, and foolproof steps. Human intervention is needed only to decide on the particular sequence or steps required to obtain the desired results. This mechanism is also known as *poka-yoka* in Total Quality Management circles. It is usually the last step toward complete automation of a process. However, just like automation, it requires a well-defined process with standard transformation steps. The verification process remains an art that, to this day, does not yield itself to well-defined steps.

Redundancy

Have two individuals check each other's work.

The final alternative to removing human errors is redundancy. It is the simplest, but also the most costly mechanism. Redundancy requires every transformation resource to be duplicated. Every transformation accomplished by a human is either independently verified by another individual, or two complete and separate transformations are performed with each outcome compared to verify that both produced the same or equivalent output. This mechanism is used in high-reliability environments, such as airborne systems. It is also used in industries where later redesign and replacement of a defective product would be more costly than the redundancy itself, such as ASIC design.

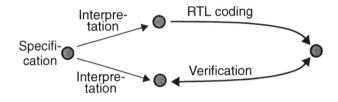

Figure 1-5. Redundancy in an ambiguous situation enables accurate verification

A different person should be in charge of verification.

Figure 1-5 shows the reconvergent paths model where redundancy is used to guard against misinterpretation of an ambiguous specification document. When used in the context of hardware design, where the transformation process is writing RTL code from a written specification document, this mechanism implies that a different individual must be in charge of the verification.

WHAT IS BEING VERIFIED?

Choosing the common origin and reconvergence points determines what is being verified. These origin and reconvergence points are often determined by the tool used to perform the verification. It is important to understand where these points lie to know which transformation is being verified. Formal verification, model checking, functional verification, and testbench generators verify different things because they have different origin and reconvergence points.

Formal Verification

Formal verification does not eliminate the need to write testbenches.

Formal verification is often misunderstood initially. Engineers unfamiliar with the formal verification process often imagine that it is a tool that mathematically determines whether their design is correct, without having to write testbenches. Once you understand what the end points of the formal verification reconvergent paths are, you know what exactly is being verified.

Formal verification falls under two broad categories: equivalence checking and model checking.

Equivalence Checking

Equivalence
checking com-
pares two models.

Figure 1-6 shows the reconvergent path model for equivalence checking. This formal verification process mathematically proves that the origin and output are logically equivalent and that the transformation preserved its functionality.

Figure 1-6.
Equivalence
checking paths

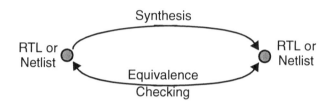

It can compare two
netlists.

In its most common use, equivalence checking compares two netlists to ensure that some netlist post-processing, such as scan-chain insertion, clock-tree synthesis, or manual modification[1], did not change the functionality of the circuit.

It can detect bugs
in the synthesis
software.

Another popular use is to verify that the netlist correctly implements the original RTL code. If one trusted the synthesis tool completely this verification would not be necessary. However, synthesis tools are large software systems that depend on the correctness of algorithms and library information. History has shown that such systems are prone to error. Equivalence checking is used to keep the synthesis tool honest. In some rare instances, this form of equivalence checking is used to verify that manually written RTL code faithfully represents a legacy gate-level design.

Less frequently, equivalence checking is used to verify that two RTL descriptions are logically identical, sometimes to avoid running lengthy regression simulations when only minor non-functional changes are made to the source code to obtain better synthesis results.

Equivalence
checking found a
bug in an arith-
metic operator.

Equivalence checking is a true alternative path to the logic synthesis transformation being verified. It is only interested in comparing boolean and sequential logic functions, not mapping these functions to a specific technology while meeting stringent design constraints.

---

1. *vi* and *emacs* remain the greatest design tools!

Engineers using equivalence checking found a design at Digital Equipment Corporation to be synthesized incorrectly. The design used a synthetic operator that was functionally incorrect when handling more than 48 bits. To the synthesis tool's defense, the documentation of the operator clearly stated that correctness was not guaranteed above 48 bits. Since the synthesis tool had no knowledge of documentation, it could not know it was generating invalid logic. Equivalence checking quickly identified a problem that could have been very difficult to detect using gate-level simulation.

Model Checking

Model checking looks for generic problems or violation of user-defined rules about the behavior of the design.

Model checking is a more recent application of formal verification technology. In it, assertions or characteristics of a design are formally proven or disproved. For example, all state machines in a design could be checked for unreachable or isolated states. A more powerful model checker may be able to determine if deadlock conditions can occur.

Another type of assertion that can be formally verified relates to interfaces. Using a formal description language, assertions about the interface are stated and the tool attempts to prove or disprove them. For example, an assertion might state that, given that signal ALE will be asserted, then either the DTACK or ABORT signal will be asserted eventually.

Figure 1-7.
Model checking paths

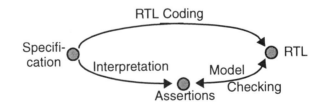

The reconvergent path model for model checking is shown in Figure 1-7 . The greatest obstacle for model checking technology is identifying, through interpretation of the design specification, which assertions to prove. Of those assertions, only a subset can feasibly be proven. Current technology cannot prove high-level assertions about a design to ensure that complex functionality is correctly implemented. It would be nice to be able to assert that, given specific register settings, a set of Asynchronous Transfer

Knowing which assertions to prove and expressing them correctly is the most difficult part.

Mode (ATM) cells will end up at a set of outputs in some relative order. Unfortunately, model checking technology is not at that level yet.

Functional Verification

Figure 1-8.
Functional verification paths

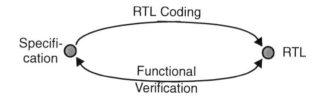

Functional verification verifies design intent.

The main purpose of functional verification is to ensure that a design implements intended functionality. As shown by the reconvergent path model in Figure 1-8, functional verification reconciles a design with its specification. Without functional verification, one must trust that the transformation of a specification document into RTL code was performed correctly, without misinterpretation of the specification's intent.

You can prove the presence of bugs, but you cannot prove their absence.

It is important to note that, unless a specification is written in a formal language with precise semantics,[2] it is <u>impossible</u> to prove that a design meets the intent of its specification. Specification documents are written using natural languages by individuals with varying degrees of ability in communicating their intentions. Any document is open to interpretation. Functional verification, as a process, can *show* that a design meets the intent of its specification. But it cannot *prove* it. One can easily prove that the design does <u>not</u> implement the intended function by identifying a single discrepancy. The converse, sadly, is not true: no one can prove that there are *no* discrepancies, just as no one can prove that flying reindeers or UFOs do not exist. (However, producing a single flying reindeer or UFO would be sufficient to prove the opposite!)

---

2. Even if such a language existed, one would eventually have to show that this description is indeed an accurate description of the design intent, based on some higher-level ambiguous specification.

Testbench Generation

Tools can generate stimulus to exercise code or expose bugs.

The increasing popularity of code coverage (see section titled "Code Coverage" on page 40 for more details) and model checking tools has created a niche for a new breed of verification tools: testbench generators. Using the code coverage metrics or the results of some proof, and the source code under analysis, testbench generators generate testbenches to either increase code coverage or to exercise the design to violate a property.

Designer input is still required.

These tools appear attractive on the surface, but, as shown in Figure 1-9, do little to contribute to the verification process. The RTL code is the common origin and there is no reconvergence point. The verification engineer is left to determine if the testbench applies valid stimulus to the design. In the affirmative, he or she must then determine, based on this stimulus, what the expected output is and compare it to the output that was produced by the design.

Figure 1-9.
Testbench
generator
paths

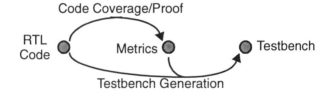

The jury is still out on the usefulness of these tools.

The usefulness of testbenches generated from code coverage metrics will be discussed in the next chapter (see section titled "What Does 100 Percent Coverage Mean?" on page 45). Testbenches generated from model checking results are useful only to illustrate how a property can be violated and what input sequence leads to the improper behavior. It may be a useful vehicle for identifying pathological conditions that were not taken into account in the specification or to provide a debugging environment for fixing the problem.

FUNCTIONAL VERIFICATION APPROACHES

Functional verification can be accomplished using three complementary but different approaches: black-box, white-box, and grey-box.

Black-Box Verification

Black-box verification cannot look at or know about the inside of a design.

With a black-box approach, the functional verification must be performed without any knowledge of the actual implementation of a design. All verification must be accomplished through the available interfaces, without direct access to the internal state of the design, without knowledge of its structure and implementation. This method suffers from an obvious lack of visibility and controllability. It is often difficult to set up an interesting state combination or to isolate some functionality. It is equally difficult to observe the response from the input and locate the source of the problem. This difficulty arises from the frequently long delays between the occurrence of a problem and the apparition of its symptom on the design's outputs.

Testcase is independent of implementation.

The advantage of black-box verification is that it does not depend on any specific implementation. Whether the design is implemented in a single ASIC, multiple FPGAs, a circuit board, or entirely in software, is irrelevant. A black-box functional verification approach forms a true conformance verification that can be used to show that a particular design implements the intent of a specification regardless of its implementation.

In black-box verification, it is difficult to control and observe specific features.

In very large or complex designs, black-box verification requires some non-functional modifications to provide additional visibility and controllability. Examples include additional software-accessible registers to control or observe internal states, or modify the size of the processed data to minimize verification time. These registers would not be used during normal operations. They are often valuable during the integration phase of the first prototype systems.

The black-box approach is the only one that can be used if the functional verification is to be implemented in parallel with the implementation of the design itself. Because there is no implementation to know about beforehand, black-box verification is the only possible avenue.

My mother is a veteran of the black-box approach: to prevent us from guessing the contents of our Christmas gifts, she never puts any names on the wrapped boxes. At Christmas, she has to correctly identify the content of each box, without opening it, so it can be given to the intended recipient. She has been known to fail on a few occasions, to the pleasure of the rest of the party!

White-Box Verification

White box verification has intimate knowledge and control of the internals of a design.

As the name suggests, a white-box approach has full visibility and controllability of the internal structure and implementation of the design being verified. This method has the advantage of being able to quickly set up an interesting combination of states and inputs, or isolate a particular function. It can then easily observe the results as the verification progresses and immediately report any discrepancies from the expected behavior.

White-box verification is tied to a specific implementation.

However, this approach is tightly integrated with a particular implementation and cannot be used on alternative implementations or future redesigns. It also requires detailed knowledge of the design implementation to know which significant conditions to create and which results to observe.

White-box verification is a useful complement to black-box verification. This approach can ensure that implementation-specific features behave properly, such as counters rolling over after reaching their end count value, or datapaths being appropriately steered and sequenced.

Grey-Box Verification

Grey box verification is a black box testcase written with full knowledge of internal details.

Grey-box verification is a compromise between the aloofness of a black-box verification and the dependence on the implementation of white-box verification. The former may not fully exercise all parts of a design, while the latter is not portable.

Testcase may not be relevant on another implementation.

As in black-box verification, a grey-box approach controls and observes a design entirely through its top-level interfaces. However, the particular verification being accomplished is intended to exercise significant features specific to the implementation. The same verification on a different implementation would be successful but may not be particularly more interesting than any other black-box verification.

TESTING VERSUS VERIFICATION

Testing verifies manufacturing.

Testing is often confused with verification. The purpose of the former is to verify that the design was manufactured correctly. The purpose of the latter is to ensure that a design meets its functional intent.

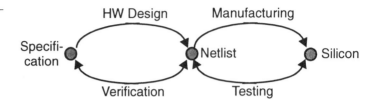

Figure 1-10.
Testing vs.
Verification

Figure 1-10 shows the reconvergent paths models for both verification and testing. During testing, the finished silicon is reconciled with the netlist that was submitted for manufacturing.

Testing verifies that internal nodes can be toggled.

Testing is accomplished through test vectors. The objective of these test vectors is not to exercise functions. It is to exercise physical locations in the design to ensure that they can go from 0 to 1 and from 1 to 0. The ratio of physical locations tested to the total number of such locations is called *test coverage*. The test vectors are usually automatically generated to maximize coverage while minimizing vectors through a process called *Automatic Test Pattern Generation* (ATPG).

Thoroughness of testing depends on controllability and observability of internal nodes.

Testing and test coverage depends on the ability to set internal physical locations to either 1 or 0, and then observe that they were indeed appropriately set. Some designs have very few inputs and output, but have a large number of possible states, requiring long sequences to properly observe and control all internal physical locations. A perfect example is an electronic wrist-watch: its has three or four inputs (the buttons around the dial) and a handful of outputs (the digits and symbols on the display). However, if it includes a calendar function, it has billions of possible states (milliseconds in hundreds of years). At speed, it would take hundreds of years to take such a design through all of its possible states.

Scan-Based Testing

Linking all registers into a long shift register increases controllability and observability.

Fortunately, scan-based testability techniques help reduce this problem to something manageable. With scan-based tests, all registers inside a design are hooked-up in a long serial chain. In normal mode, the registers operate as if the scan chain was not there (see

Figure 1-11(a)). In scan mode, the registers operate as a long shift register (see Figure 1-11(b)).

To test a scannable design, the unit under test is put into scan mode, then an input pattern is shifted through all of its internal registers. The design is then put into normal mode and a <u>single</u> clock cycle is applied, loading the result of the normal operation based on the scanned state into the registers. The design is then put into scan mode again. The result is shifted out of the registers (at the same time the next input pattern is shifted in) and the result is compared against the expected value.

Figure 1-11.
Scan-based
testing

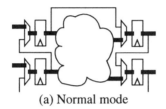 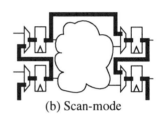

(a) Normal mode   (b) Scan-mode

Scan-based test-
ing puts restric-
tions on design.

This increase in controllability and observability, and thus test coverage, comes at a cost. Certain restrictions are put onto the design to enable the insertion of a scan chain and the automatic generation of test patterns. Some of these restrictions include, but are not limited to: fully synchronous design, no derived or gated clocks, and use of a single clock edge. The topic of design for testability is far greater and complex than this simple introduction implies. For more details, there are several excellent books[3] and papers[4] on the subject.

But the drawbacks
of these restric-
tions are far out-
weighed by the
benefits of scan-
based testing.

Hardware designers introduced to scan-based testing initially rebel against the restrictions imposed on them. They see only the immediate area penalty and their favorite design technique rendered inadequate. However, the increased area and additional design effort are quickly outweighed when a design can be fitted with one or more scan chains, when test patterns are generated and high test

3. Abramovici, Breuer, and Friedman. *Digital System Testing and Testable Design.* IEEE. ISBN 0780310624

4. Cheung and Wang, *"The Seven Deadly Sins of Scan-Based Design"*, Integrated System Design, Aug. 1997, p50-56.

coverage is achieved automatically, at the push of a button. The time saved and the greater confidence in putting a working product on the market far outweighs the added cost for scan-based design.

Design for Verification

Design practices need to be modified to accommodate testability requirements. Isn't it acceptable to modify those same design practices to accommodate verification requirements?

Verification must be considered during specification.

With functional verification taking twice as much effort as the design itself, it is reasonable to require additional design effort to simplify verification. Just as scan chains are put in a design to improve testability without adding to the functionality, it should be standard practice to add non-functional structures and features to facilitate verification. This requires that verification be considered at the outset of a project, during its specification phase. Not only should the architect of the design answer the question "*what is this supposed to do?*", but also "*how is this thing going to be verified?*"

Typical design-for-verification techniques include providing additional software-accessible registers to control and observe internal locations, and providing programmable multiplexors to isolate or by pass functional units.

VERIFICATION AND DESIGN REUSE

Today, design reuse is considered the best way to overcome the difference between the number of transistors that can be manufactured on a single chip, and the number of transistors engineers can take advantage of in a reasonable amount of time. This difference is called the *productivity gap*. Design reuse is a simple concept that seems easy to put in practice. The reality is proving to be more problematic.

Reuse Is About Trust

You won't use what you do not trust.

The major obstacle to design reuse is cultural. Engineers have little incentive and willingness to incorporate an unknown design into their own. They do not trust that the other design is as good or as reliable as one designed by themselves. The key to design reuse is gaining that trust.

Trust, like quality, is not something that can be added to a design after the fact. It must be built-in, through the best possible design practices. And it must be earned by standing behind a reusable design: providing support services and building a relationship with the user. Once that trust is established, reuse will happen more often.

Proper functional verification demonstrates trustworthiness of a design.

Trustworthiness can also be demonstrated through a proper verification process. By showing the user that a design has been thoroughly and meticulously verified, according to the design specification, trust can be built and communicated much faster. Functional verification is the only way to demonstrate that the design meets, or even exceeds, the quality of a similar design that an engineer could do himself or herself.

Verification for Reuse

Reusable designs must be verified to a greater degree of confidence.

If you create a design, you have a certain degree of confidence in your own abilities as a designer and implicitly trust its correctness. Functional verification is used only to confirm that opinion and to augment that opinion in areas known to be weak. If you try to reuse a design, you can rely only on the functional verification to build that same level of confidence and trust. Thus, reusable designs must be verified to a greater degree of confidence than custom designs.

All claims, possible configurations and uses must be verified.

Because reusable designs tend to be configurable and programmable to meet a variety of possible environment and applications, it is necessary to verify a reusable design under all possible configurations and for all possible uses. All claims made about the reusable design must be verified and demonstrated to users.

THE COST OF VERIFICATION

Verification is a necessary evil. It always takes too long and costs too much. Verification does not generate a profit or make money: after all, it is the design being verified that will be sold and ultimately make money, not the verification. Yet verification is indispensable. To be marketable and create revenues, a design must be functionally correct and provide the benefits that the customer requires.

As the number of errors left to be found decreases, the time - and cost - to identify them increases.

Verification is a process that is never truly complete. The objective of verification is to ensure that a design is error-free, yet one cannot prove that a design is error-free. Verification can only show the *presence* of errors, not their *absence*. Given enough time, an error <u>will</u> be found. The question thus becomes: is the error likely to be severe enough to warrant the effort spent identifying it? As more and more time is spent on verification, fewer and fewer errors are found with a constant effort expenditure. As verification progresses, it has diminishing returns. It costs more and more to find fewer and fewer, often unlikely, errors.

Functional verification is similar to statistical hypothesis testing. The hypothesis under test is: *is my design functionally correct?* The answer can be either yes or no. But either answer could be wrong. These wrong answers are Type II and Type I mistakes, respectively.

Figure 1-12. Type I & II mistakes

	Errors	No Errors
Bad Design		Type II (False Positive)
Good Design	Type I (False Negative)	

False positives must be avoided.

Figure 1-12 shows where each type of mistake occurs. Type I mistakes, or *false negatives*, are the easy ones to identify. The verification is finding an error where none exist. Once the misinterpretation is identified, the implementation of the verification is modified to change the answer from **no** to **yes**, and the mistake no longer exists. Type II mistakes are the most serious ones: the verification failed to identify an error. In a Type II mistake, or *false positive* situation, a bad design is shipped unknowingly, with all the potential consequences that entails.

The United States Food and Drug Administration faces Type II mistakes on a regular basis with potentially devastating consequences: in spite of positive clinical test results, is a dangerous drug released on the market? Shipping a bad design may result in simple product recall or in the total failure of a space probe after it has landed on another planet.

With the future of the company potentially at stake, the 64-thousand dollar question in verification is: *how much is enough*? The functional verification process presented in this book, along with some of the tools described in the next chapter attempt to answer that question.

The 64-million dollar question is: *when will I be done*? Knowing where you are in the verification process, although impossible to establish with certainty, is much easier to estimate than how long it will take to complete the job. The verification planning process described in Chapter 3 creates a tool that enables a verification manager to better estimate the effort and time required to complete the task at hand, to the degree of certainty required.

SUMMARY

In this chapter, after briefly outlining what a testbench is, I described the importance of verification. Parallelism, automation and abstractions were identified as strategies to reduce the time necessary to implement testbenches. A model was developed to illustrate and identify what exactly is being verified in any verification process. It was then applied to various verification tools and methodologies currently available and to differentiate verification from manufacturing test. Techniques for eliminating the uncertainty introduced by human intervention in the verification process were also described. The importance of verification for design reuse and the cost of verification were also discussed.

CHAPTER 2   VERIFICATION TOOLS

As mentioned in the previous chapter, one of the mechanisms that can be used to improve the efficiency and reliability of a process is automation. This chapter covers tools used in a state-of-the-art functional verification environment. Some of these tools, such as simulators, are essential for the functional verification activity to take place. Others, such as linting or code coverage tools, automate some of the most tedious tasks of verification and help increase the confidence in the outcome of the functional verification.

Not all tools are mentioned in this chapter. It is not necessary to use all the tools mentioned.

It is not necessary to use all of the tools described here. Nor is this list exhaustive, as new application-specific and general purpose verification automation tools are made available. As a verification engineer, your job is to use the necessary tools to ensure that the outcome of the verification process is not a Type II mistake, which is a false positive. As a project manager responsible for the delivery of a working product on schedule and within the allocated budget, your responsibility is to arm your engineers with the proper tools to do their job efficiently and with the greatest degree of confidence. Your job is also to decide when the cost of finding the next functional bug have increased above the value the additional functional correctness brings. This last responsibility is the heaviest of them all. Some of these tools provide information to help you decide when you've reached that point.

No endorsements of commercial tools.

I mention some commercial tools by name. They are used for illustrative purposes only and this does not constitute a personal endorsement. I apologize in advance to suppliers of competitive products I fail to mention. It is not an indication of inferiority, but rather an indication of my limited knowledge. All trademarks and service marks, registered or not, are the property of their respective owners.

LINTING TOOLS

Linting tools find common program- mer mistakes.

The term *lint* comes from the name of a UNIX utility that parses a C program and reports questionable uses and potential problems. When the C programming language was created by Dennis Ritchie, it did not include many of the safeguards that have evolved in later versions of the language, like ANSI-C or C++, or other strongly-typed languages such as Pascal or ADA. *lint* evolved as a tool to identify common mistakes programmers made, allowing them to find the mistakes quickly and efficiently, instead of waiting to find them through a dreaded segmentation fault during verification of the program.

lint identifies real problems, such as mismatched types between arguments and function calls or mismatched number of arguments, as shown in Sample 2-1. The source code is syntactically correct and compiles without a single error or warning using *gcc* version 2.8.1.

Sample 2-1. Syntactically correct K&R C source code

```
int my_func(addr_ptr, ratio)
int* addr_ptr;
float ratio;
{
    return (*addr_ptr)++;
}

main()
{
    int my_addr;
    my_func(my_addr);
}
```

However, Sample 2-1 suffers from several pathologically severe problems:

1. The *my_func* function is called with only one argument instead of two.

2. The *my_func* function is called with an integer value as a first argument instead of a pointer to an integer value.

Problems are found faster than at run-time.

As shown in Sample 2-2, the *lint* program identifies these problems, letting the programmer fix them before executing the program and observing a catastrophic failure. Diagnosing the problems at run-time would require a run-time debugger and would take several minutes. Compared to the few seconds it took using *lint*, it is easy to see that the latter method is more efficient.

Sample 2-2.
Lint output for
Sample 2-1

```
src.c(3): warning: argument ratio unused in
function my_func
src.c(11): warning: addr may be used before set
src.c(12): warning: main() returns random value
to invocation environment
my_func: variable # of args.    src.c(4)  ::
src.c(11)
my_func, arg. 1 used inconsistently
src.c(4)  ::  src.c(11)
my_func returns value which is always ignored
```

Linting tools are static tools.

Linting tools have a tremendous advantage over other verification tools: they do not require stimulus, nor do they require a description of the expected output. They perform checks that are entirely static in nature, with the expectations built into the linting tool itself.

The Limitations of Linting Tools

Linting tools can only identify a certain class of problems.

Other potential problems were also identified by *lint*. All were fixed in Sample 2-3 but *lint* continues to report a problem with the invocation of the *my_func* function: the return value is always ignored. Linting tools cannot identify all problems in source code. They can only find problems that can be statically deduced by looking at the code structure, not problems in the algorithm or data flow. For example, in Sample 2-3, *lint* does not recognize that the uninitialized *my_addr* variable will be incremented in the *my_func* function, producing random results. Linting tools are similar to spell checkers; they identify misspelled words, but do not determine if the wrong word is used. For example, this book could have several instances of the word "with" being used instead of "width". It is a type of error the spell checker (or a linting tool) could not find.

Many false negatives are reported.

Another limitation of linting tools is that they are often too paranoid in reporting problems they identify. To avoid making a Type II mis-

Sample 2-3.
Functionally
correct K&R
C source code

```
int my_func(addr_ptr)
int* addr_ptr;
{
    return (*addr_ptr)++;
}

main()
{
    int my_addr;
    my_func(&my_addr);
    return 0;
}
```

take - reporting a false positive, they err on the side of caution and report potential problems where none exist. This results in many Type I mistakes - or false negatives. Designers can become frustrated while looking for non-existent problems and may abandon using linting tools altogether.

Carefully filter
error messages!

You should filter the output of linting tools to eliminate warnings or errors known to be false. Filtering error messages helps reduce the frustration of looking for non-existent problems. More importantly, it reduces the output clutter, reducing the probability that the report of a real problem goes unnoticed among dozens of false reports. Similarly, errors known to be true positive should be highlighted. Extreme caution must be exercised when writing such a filter: you must make sure that a true problem does not get filtered out and never reported.

Naming conven-
tions can help out-
put filtering.

A properly defined naming convention is a useful tool to help determine if a warning is significant. For example, the report in Sample 2-4 about a latch being inferred on a signal whose name ends with "_LT" would be considered as expected and a false warning. All other instances would be flagged as true errors.

Sample 2-4.
Output from a
hypothetical
Verilog lint-
ing tool

```
Warning: file decoder.v, line 23: Latch
inferred on reg "ADDRESS_LT".
Warning: file decoder.v, line 36: Latch
inferred on reg "NEXT_STATE".
```

Do not turn off
checks.

Filtering the output of a linting tool is preferable to turning off checks from within the source code itself or via the command line. A check may remain turned off for an unexpected duration, potentially hiding real problems. Checks that were thought to be irrelevant may become critical as new source files are added.

Lint code as it is being written.

Because it is better to fix problems when they are created, you should run lint on the source code while it is being written. If you wait until a large amount of code is written before linting it, the large number of reports - many of them false - will be daunting and create the impression of a setback. The best time to identify a report as true or false is when you are still intimately familiar with the code.

Enforce coding guidelines.

The linting process can also be used to enforce coding guidelines and naming conventions[1]. Therefore, it should be an integral part of the authoring process to make sure your code meets the standards of readability and maintainability demanded by your audience.

Linting Verilog Source Code

Verilog is a type-less language.

Many people compare Verilog to C and VHDL to ADA. I do not think it is a fair comparison: C (at least ANSI-C) is a strongly-typed language. In C, you cannot assign an integer value to a real variable nor vice-versa. You can in Verilog. Verilog is more like assembler or BASIC: it is a _typeless_ language. You can assign any value to any register or subprogram argument, sometimes with disastrous consequences (such as assigning the value of an expression in a narrow register, clipping the most significant bits).

Linting Verilog source code catches common errors.

Linting Verilog source code ensures that all data is properly handled without accidentally dropping or adding to it. The code in Sample 2-5 shows a Verilog model that looks perfect, compiles without errors, but produces unintended results under some circumstances.

Problems may not be apparent under most conditions.

The problem is in the width mismatch in the continuous assignment between the output "_out_" and the constant "'bz". The unsized constant is 32-bit wide (or a value of "32'hzzzzzzzz"), while the output has a user-specified width. As long as the width of the output is less than or equal to 32, everything is fine: the value of the constant will be appropriately truncated to fit the width of the output. However, the problem occurs when the width of the output is greater than 32

---

1. See Appendix A for a set of coding guidelines.

Sample 2-5.
Potentially
problematic
Verilog code

```
module tristate_buffer(in, out, enable);
parameter WIDTH = 8;
input   [WIDTH-1:0] in;
output  [WIDTH-1:0] out;
input               enable;

assign out = (enable) ? in : 'bz;

endmodule
```

bits: Verilog <u>zero-extends</u> the constant value to match the width of the output, producing the wrong result. The least significant bits is set to high-impedance while all the other more significant bits are set to zero.

It is an error that could not be found in simulation, unless a configuration greater then 32 bits was used <u>and</u> it produced wrong results at a time and place you were looking at. A linting tool finds the problem every time, in just a few seconds.

Linting VHDL Source Code

Because of its strong typing, VHDL does not need linting as much as Verilog. However, potential problems are still best identified using a linting tool.

Linting can find
unintended multi-
ple drivers.

For example, a common problem in VHDL is created by using the STD_LOGIC type. Since it is a resolved type, STD_LOGIC signals can have more than one driver. When modeling hardware, multiple driven signals are required in a single case: to model buses. In all other cases (which is over 99 percent of the time), a signal should have only one driver. The VHDL source shown in Sample 2-6 demonstrates how a simple typographical error can easily go undetected and satisfy the usually paranoid VHDL compiler.

Typographical
errors can cause
serious problems.

In Sample 2-6, both concurrent signal assignments labelled *"statement1"* and *"statement2"* assign to the signal *"s1"* (ess-one), while the signal *"sl"* (ess-ell) remains unassigned. Had I used the STD_ULOGIC type instead of the STD_LOGIC type, the VHDL toolset would have reported an error after finding multiple drivers on an unresolved signal. However, it is not possible to guarantee the STD_ULOGIC type is used for all signals with a single driver. A

Sample 2-6.
Erroneous
multiple
drivers

```
library ieee;
use ieee.std_logic_1164.all;
entity my_entity is
    port (my_input: in std_logic);
end my_entity;


architecture sample of my_entity is
    signal s1: std_logic;
    signal sl: std_logic;
begin
    statement1: s1 <= my_input;
    statement2: s1 <= not my_input;
end sample;
```

linting tool is still required to report multiply driven signals regardless of the type, as shown in Sample 2-7.

Sample 2-7.
Output from a
hypothetical
VHDL linting
tool

```
Warning: file my_entity.vhd: Signal "s1" is
multiply driven.
Warning: file my_entity.vhd: Signal "sl" has no
drivers.
```

Use naming convention to filter output.

It would be up to the author to identify the signals that were intended to model buses and ignore the warnings about them. Using a naming convention for such signals facilitates recognizing warnings that can be safely ignored, and enhances the reliability of your code. An example of a naming convention, illustrated in Sample 2-8, would be to name any signals modeling buses with the "_bus" prefix[2].

Sample 2-8.
Naming convention for
signals with
multiple drivers

```
--
-- data_bus, addr_bus and sys_err_bus
-- are intended to be multiply driven
--
signal data_bus : std_logic_vector(15 downto 0);
signal addr_bus : std_logic_vector( 7 downto 0);
signal ltch_addr: std_logic_vector( 7 downto 0);
signal sys_err_bus: std_logic;
signal bus_grant  : std_logic;
```

2. See Appendix A for an example of naming guidelines.

Linting can iden-
tify inferred
latches.

The accidental multiple driver problem is not the only one that can be caught using a linting tool. Others, such as unintended latch inference in synthesizeable code, or the enforcement of coding guidelines, can also be identified.

Code Reviews

Reviews are per-
formed by peers.

Although not technically linting tools, the objective of code reviews is essentially the same: identify functional and coding style errors before functional verification and simulation. In code reviews, the source code produced by a designer is reviewed by one or more peers. The goal is not to publicly ridicule the author, but to identify problems with the original code that could not be found by an automated tool.

Identify qualitative
problems and
functional errors.

A code review is an excellent venue for evaluating the maintainability of a source file, and the relevance of its comments. Other qualitative coding style issues can also be identified. If the code is well understood, it is often possible to identify functional errors or omissions.

Code reviews are not new ideas either. They have been used for many years in the software design industry. Detailed information on how to conduct effective code reviews can be found in the *resources* section at:

http://janick.bergeron.com/wtb

SIMULATORS

Simulate your
design before
implementing it.

Simulators are the most common and familiar verification tools. They are named *simulators* because their role is limited to approximating reality. A simulation is never the final goal of a project. The goal of all hardware design projects is to create real physical designs that can be sold and generate profits. Simulators attempt to create an artificial universe that mimics the future real design. This lets the designers interact with the design before it is manufactured and correct flaws and problems earlier.

Simulators are
only approxima-
tions of reality.

You must never forget that a simulator is an approximation of reality. Many physical characteristics are simplified - or even ignored - to ease the simulation task. For example, a digital simulator assumes that the only possible values for a signal are '0', '1', unknown, and high-impedance. However, in the physical - and analog - world, the value of a signal is a continuous function of the voltage and current across a thin aluminium or copper wire track: an infinite number of possible values. In a discrete simulator, events that happen deterministically 5 ns apart may be asynchronous in the real world and may occur randomly.

Simulators are at the mercy of the descriptions being simulated.

Within that simplified universe, the only thing a simulator does is execute a description of the design. The description is limited to a well-defined language with precise semantics. If that description does not accurately reflect the reality it is trying to model, there is no way for you to know that you are simulating something that is different from the design that will be ultimately manufactured. Functional correctness and accuracy of models is a big problem as errors cannot be proven *not* to exist.

Stimulus and Response

Simulation requires stimulus.

Simulators are not static tools. A static verification tool performs its task on the design without any additional information or action required by the user. For example, linting tools are static tools. Simulators, on the other hand, require that you provide a facsimile of the environment in which the design will find itself. This facsimile is often called a testbench. Writing this testbench is the main objective of this textbook. The testbench needs to provide a representation of the inputs observed by the design, so the simulator can emulate the design's responses based on its description.

The simulation outputs are validated externally, against design intents.

The other thing that you must not forget is that simulators have no knowledge of your intentions. They cannot determine if a design being simulated is correct. Correctness is a value judgment on the outcome of a simulation that must be made by you, the designer. Once the design is submitted to an approximation of the inputs from its environment, your primary responsibility is to examine the outputs produced by the simulation of the design's description and determine if that response is appropriate.

Event-Driven Simulation

Simulators are never fast enough.

Simulators are continuously faced with one intractable problem: they are never fast enough. They are attempting to emulate a physical world where electricity travels at the speed of light and transistors switch over one billion times in a second. Simulators are implemented using general purpose computers that can execute, under ideal conditions, up to 100 million instructions per second. The speed advantage is unfairly and forever tipped in favor of the physical world.

Outputs change only when an input changes.

One way to optimize the performance of a simulator is to avoid simulating something that does not need to be simulated. Figure 2-1 shows a 2-input XOR gate. In the physical world, if the inputs do not change (Figure 2-1(a)), even though voltage is constantly applied to the output, current is continuously flowing through the transistors (in some technologies), and the atomic particles in the semiconductor are constantly moving, the *interpretation* of the output electrical state as a binary value (either a logic '1' or a logic '0') does not change. Only if one of the inputs change (as in Figure 2-1(b)) does the output change.

Figure 2-1.
Behavior of an
XOR gate

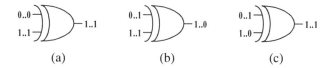

(a) (b) (c)

Change in values, called *events*, drive the simulation process.

Sample 2-9 shows a VHDL description (or model) of an XOR gate. The simulator could choose to continuously execute this model, producing the same output value if the input values did not change. An opportunity to improve upon that simulator's performance becomes obvious: do not execute the model while the inputs are constants. Phrased another way: only execute a model when an input changes. The simulation is therefore driven by changes in inputs. If you define an input change as an *event*, you now have an *event-driven* simulator.

Sample 2-9.
VHDL model
for an XOR
gate

```
XOR_GATE: process (A, B)
begin
    if A = B then
        O <= '0';
    else
        O <= '1'
    end if;
end process XOR_GATE;
```

Sometimes, input changes do not cause the output to change.

But what if both inputs change, as in Figure 2-1(c)? In the physical world, the output does not change. What should an event-driven simulator do? For two reasons, the simulator should execute the description of the XOR gate. First, in the real world, the output of the XOR gate *does* change. The output might oscillate between '0' and '1' or remain in the "neither '0' nor '1'" region for a few hundredths of picoseconds (see Figure 2-2). It just depends on how accurate you want your model to be. You could decide to model the XOR gate to include the small amount of time spent in the unknown (or 'x') state to more accurately reflect what happens when both inputs change at the same time.

Figure 2-2. Behavior of an XOR gate when both inputs change

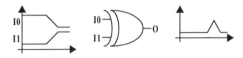

Descriptions between inputs and outputs are arbitrary.

The second reason is that the event-driven simulator does not know apriori that it is about to execute a model of an XOR gate. All the simulator knows is that it is about to execute a description of a 2-input, 1-output function. Figure 2-3 shows the view of the XOR gate from the simulator's perspective: a simple 2-input, 1-output black box. The black box could just as easily contain a 2-input AND gate (in which case the output might very well change if both inputs change), or a 1024-bit linear feedback shift register (LFSR).

Figure 2-3. Event-driven simulator view of an XOR gate

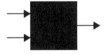

The mechanism of event-driven simulation introduces some limitations and interesting side effects that are discussed further in Chapter 4.

Cycle-Based Simulation

Figure 2-4 shows the event-driven view of a synchronous circuit composed of a chain of three two-input gates between two edge-triggered flip-flops. Assuming that all other inputs remain constant, a rising edge on the clock input would cause an event-driven simulator to simulate the circuit as follows:

Figure 2-4.
Event-driven
simulator view
of a
synchronous
circuit

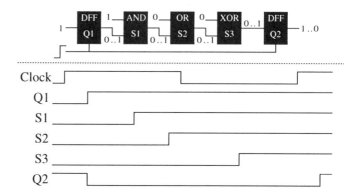

1. The event (rising edge) on the clock input causes the execution of the description of the flip-flop models, changing the output value of Q1 to '1' and of Q2 to '0', after a delay of 1 ns.

2. The event on Q1 causes the description of the AND gate to execute, changing the output S1 to '1', after a delay of 2 ns.

3. The event on S1 causes the description of the OR gate to execute, changing the output S2 to '1', after a delay of 1.5 ns.

4. The event on S2 causes the description of the XOR gate to execute, changing the output S3 to '1' after a delay of 3 ns.

5. The next rising edge on the clock causes the description of the flip-flops to execute, Q1 remaining unchanged but Q2 changing back to '1', after a delay of 1 ns.

Many intermediate events in synchronous circuits are not functionally relevant.

To simulate the effect of a single clock cycle on this simple circuit required the generation of six events and the execution of seven models. If all we are interested in are the final states of Q1 and Q2, not of the intermediate combinatorial signals, the simulation of this circuit could be optimized by acting only on the significant events for Q1 and Q2: the active edge of the clock. Phrased another way: simulation is based on clock cycles. This is how cycle-based simulators operate.

The synchronous circuit in Figure 2-4 can be simulated in a cycle-based simulator using the following sequence:

Writing Testbenches: Functional Verification of HDL Models

Cycle-based simulators collapse combinatorial logic into equations.

1. When the circuit description is compiled, all combinatorial functions are collapsed into a single expression that can be used to determine all flip-flop input values based on the current state of the fan-in flip-flops.

   For example, the combinatorial function between Q1 and Q2 would be compiled from the following initial description:

   ```
   S1 = Q1 & '1'
   S2 = S1 | '0'
   S3 = S2 ^ '0'
   ```

   into this final single expression

   ```
   S3 = Q1
   ```

   The cycle-based simulation view of the compiled circuit is shown in Figure 2-5.

Figure 2-5.
Cycle-based simulator view of a synchronous circuit

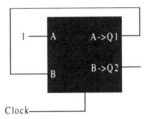

2. During simulation, whenever the clock input rises, the value of all flip-flops are updated using the input value returned by the pre-compiled combinatorial input functions.

   The simulation of the same circuit, using a cycle-based simulator, required the generation of two events and the execution of a single model.

Cycle-based simulations have no timing information.

This great improvement in simulation performance comes at a cost: all timing and delay information is lost. Cycle-based simulators assume that the entire design meets the set-up and hold requirements of all the flip-flops. When using a cycle-based simulator, timing is usually verified using a static timing analyzer.

Cycle-based simulators can only handle synchronous circuits.

Cycle-based simulators further assume that the active clock edge is the only significant event in changing the state of the design. All other inputs are assumed to be perfectly synchronous with the active clock edge. Therefore, cycle-based simulators can only simulate perfectly synchronous designs. Anything containing asynchronous inputs, latches, or multiple-clock domains <u>cannot</u> be simulated accurately. Fortunately, the same restrictions apply to static timing analysis. Thus, circuits that are suitable for cycle-based simulation to verify the functionality, are suitable for static timing verification to verify the timing.

Co-Simulators

Very few real-world circuits are perfectly synchronous. Some may have a single clock domain and use only flip-flops, but their external inputs are likely to be asynchronous. Few real-world circuits could thus take advantage of cycle-based simulation.

Multiple simulators can handle separate portions of a design.

To handle the portions of a design that do not meet the requirements for cycle-based simulation, most simulators are integrated with an event-driven simulator. As shown in Figure 2-6, the synchronous portion of the design is simulated using the cycle-based algorithm, while the remainder of the design is simulated using a conventional event-driven simulator. Both simulators (event-driven and cycle-based) are running together, cooperating to simulate the entire design.

Figure 2-6.
Event-driven
and cycle-
based co-
simulation

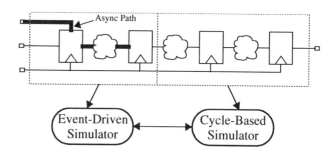

Other popular co-simulation environments provide VHDL and Verilog, HDL and C, or digital and analog co-simulation.

All simulators operate in lock-step.

During co-simulation, all simulators involved progress along the time axis in lock-step. All are at simulation time T_1 at the same time and reach the next time T_2 at the same time. This implies that the speed of a co-simulation environment is limited by the slowest simulator. Some experimental co-simulation environments implement *time warp* synchronization where some simulators are allowed to move ahead of the others.

Performance is decreased by the communication and synchronization overhead.

The biggest hurdle of co-simulation comes from the communication overhead between the simulators. Whenever a signal generated within a simulator is required as an input by another, the current value of that signal, as well as the timing information of any change in that value, must be communicated. This communication usually involves a translation of the event from one simulator into an (almost) equivalent event in another simulator. Ambiguities can arise during that translation when each simulation has different semantics. The difference in semantics is usually present: the semantic difference often being the requirement for co-simulation in the first place.

Translating values and events from one simulator to another can create ambiguities.

Examples of translation ambiguities abound. How do you map Verilog's 128 possible states (composed of orthogonal logic values and strengths) into VHDL's nine logic values (where logic values and strengths are combined)? How do you translate a voltage and current value in an analog simulator into a logic value and strength in a digital simulator? How do you translate the timing of zero-delay events from Verilog (which has no strict concept of delta cycles)[3] to VHDL?

Co-simulation should not be confused with single-kernel simulation.

Co-simulation is when two (or more) simulators are cooperating to simulate a design, each simulating a portion of the design, as shown in Figure 2-7. It should not be confused with simulators able to read and compile models described in different languages. For example, Cadence's *NCSIM* simulator and Model Technology's *ModelSim* simulator can both simulate a design described using a mix of VHDL and Verilog. As shown in Figure 2-8, both languages are compiled into a single internal representation or machine code and the simulation is performed using a single simulation engine.

Figure 2-7.
Co-simulator

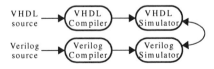

3. See "The Simulation Cycle" on page 129 for more details on delta cycles.

Figure 2-8.
Mixed-
language
simulator

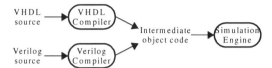

THIRD-PARTY MODELS

Board-level
designs should
also be simulated.

Your board-level design likely contains devices that were pur-chased from a third party. Sometimes these devices are programma-ble, such as memories, PLDs, or FPGAs. You should verify your board design to ensure that the ASICs interoperate properly between themselves and with the third-party components. You should also make sure that the programmable parts are functionally correct or simply verify that the connectivity, which has been hand-captured via a schematic, is correct.

You can buy mod-els for standard parts.

If you want to verify your board design, it is necessary to have models for all the parts included in a simulation. System-level sim-ulations, composed of a subset of board designs or multiple board designs, also require models for these external components. If you were able to procure the part from a third party, you should be able to procure a model of that part as well. You may have to obtain the model from a different vendor than the one who supplies the physi-cal part.

There are several providers of models for standard SSI and LSI components, memories and processors. Many are provided as non-synthesizeable VHDL or Verilog source code. For intellectual prop-erty protection and licensing technicalities, most are provided as compiled binary models.

It is cheaper to buy models than write them yourself.

At first glance, procuring a model of a standard part from a third-party provider may seem expensive. Many have decided to write their own models to save on licensing costs. However, you have to decide if this endeavor is truly economically fruitful. If you have a shortage of qualified engineers, why spend critical resources on writing a model that does not embody any competitive advantage for your company? If it was not worth designing on your own in the first place, why is writing your own model suddenly justified?

Your model is not as reliable as the one you buy.

Secondly, the model you write has never been used before. Its quality is much lower than a model that has been used by several other companies before you. The value of a functionally correct and reliable model is far greater than an uncertain one. Writing and verifying a model to the *same degree of confidence* as the third-party model is always more expensive than licensing it. And be assured: no matter how simple the model is (such as a quad 2-input NAND gate, 74LS00), you'll get it wrong the first time. If not functionally, then at least with respect to timing or connectivity.

Hardware Modelers

What if you cannot find a model to buy?

You may be faced with procuring a model for a device that is so new or so complex, that no provider has had time to develop a reliable model for it. For example, at the time this book was written, you could license full-functional models for the Pentium processor from at least two vendors. However, you could not find a model for the Pentium III. If you want to verify that your new PC board, which uses the latest Intel microprocessor, is functionally correct before you build it, you have to find some other way to include a simulation model of the processor.

You can "plug" a chip into a simulator.

Hardware modelers provide a solution for that situation. A hardware modeler is a small box that connects to your network. A <u>real</u>, <u>physical</u> chip that needs to be simulated is plugged in it. During simulation, the hardware modeler communicates with your simulator (through a special interface package) to supply inputs from the simulator to the device, then sends the sampled output values from the device back to the simulation. Figure 2-9 illustrates this communication process.

Figure 2-9. Interfacing a hardware modeler and a simulator

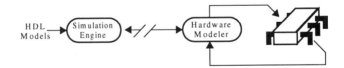

Timing of I/O signals still needs to be modeled.

Using a hardware modeler is not a trivial task. Often, an adaptor board must be built to fit the device onto the socket on the modeler itself. Also, the modeler cannot perform timing checks on the device's inputs nor accurately reflect the output delays. A timing shell performing those checks and delays must be written to more accurately model a device using a hardware modeler.

Hardware modelers offer better simulation performance.

Hardware modelers are also very useful when simulating a model of the part at the required level of abstraction. A full-functional model of a modern processor that can fetch, decode and execute instructions could not realistically execute more than 10 to 50 instructions within an acceptable time period. The real physical device can perform the same task in a few milliseconds. Using a hardware modeler can greatly speed up board- and system-level simulation.

WAVEFORM VIEWERS

Waveform viewers display the changes in signal values over time.

Waveform viewers are the most common verification tools used in conjunction with simulators. They let you visualize the transitions of multiple signals over time, and their relationship with other transitions. With such a tool, you can zoom in and out over particular time sequences, measure time differences between two transitions, or display a collection of bits as bit strings, hexadecimal or as symbolic values. Figure 2-10 shows a typical display of a waveform viewer showing the inputs and outputs of a 4-bit synchronous counter.

Figure 2-10.
Hypothetical waveform view of a 4-bit synchronous counter

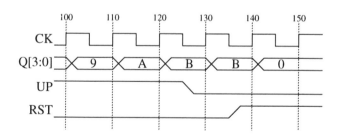

Waveform viewers are used to debug simulations.

Waveform viewers are indispensable during the authoring phase of a design or a testbench. With a viewer you can casually inspect that the behavior of the code is as expected. They are needed to diagnose, in an efficient fashion, why and when problems occur in the design or testbench. They can be used interactively during the simulation, but more importantly offline, after the simulation has completed. As shown in Figure 2-11, a waveform viewer can play back the events that occurred during the simulation that were recorded in some trace file.

Figure 2-11. Waveform viewing as post-processing

Recording waveform trace data decreases simulation performance.

Viewing waveforms as a post-processing step lets you quickly browse through a simulation that can take hours to run. However, keep in mind that recording trace information significantly reduces the performance of the simulator. The quantity and scope of the signals whose transitions are traced, as well as the duration of the trace, should be limited as much as possible.

Do not use a waveform viewer to determine if a design passes or fails.

In a functional verification environment, using a waveform viewer to determine the correctness of a design involves interpreting the dozens (if not hundreds) of wavy lines on a computer screen against some expectation. It can be an acceptable verification method used two or three times, for less than a dozen signals. But, as the number of signals increases, and the number of transitions on these signals increases, and the number of relationships between transitions critical to the correctness increases, and the duration of the simulation that must be checked for correctness increases, and the number of times simulation results must be checked increases, the probability that a functional error is missed increases exponentially and quickly reaches one.

Some viewers can compare sets of waveforms.

Some waveform viewers can compare two sets of waveforms. One set is presumed to be a golden reference, while the other is verified for any discrepancy. The comparator visually flags or highlights any differences found. This approach has two significant problems.

How do you define a set of waveforms as "golden"?

First, how is the golden reference waveform set declared "golden"? If visual inspection is required, the probability of missing a significant functional error remains equal to 1 in most cases. The only time golden waveforms are truly available is in a redesign exercise, where cycle-accurate backward compatibility must be maintained. However, there are very few of these designs. Most redesign exercises take advantage of the process to introduce needed modifications or enhancements, thus tarnishing the status of the golden waveforms.

And are the differ-
ences really signif-
icant?

Second, differences from the golden waveforms may not be significant. The value of all output signals is not significant all the time. Sometimes, what is significant is the relative relationships between the transitions, not their absolute position. The new waveforms may be simply shifted by a few clock cycles compared to the reference waveforms, but remain functionally correct. Yet, the comparator identifies this situation as a mismatch.

CODE COVERAGE

Did you forget to
verify some func-
tion in your code?

Code coverage is a methodology that has been in use in software engineering for quite some time. The problem with false positive answers (i.e. a bad design is thought to be good), is that they look identical to a true positive answer. It is impossible to know, with 100 percent certainty, that the design being verified is indeed functionally correct. All of your testbenches simulate successfully, but is there a function or a combination of functions that you forgot to verify? That is the question that code coverage can help answer.

Code must first be
instrumented.

Figure 2-12 shows how a code coverage tool works. The source code is first *instrumented*. The instrumentation process simply adds checkpoints at strategic locations of the source code to record whether a particular construct has been exercised. The instrumentation method varies from tool to tool. Some may use file I/O features available in the language (i.e. use *$write* statements in Verilog or *textio.write* procedure calls in VHDL). Others may use special features built into the simulator.

Figure 2-12.
Code coverage
process

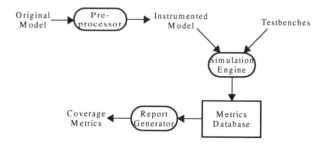

No need to instrument the testbenches.

Only the code for the design under test is instrumented. The objective is to determine if you have forgotten to exercise some code in the design. The code for the testbenches need not be traced to confirm that is has executed. If a significant section of a testbench was not executed, it should be reflected in some portion of the design not being exercised. If not, that section may not be as significant as once thought.

Trace information is collected at runtime.

The instrumented code is then simulated normally using all available, uninstrumented, testbenches. The cumulative traces from all simulations are collected into a database. From that database, reports can be generated to determine various coverage metrics of the verification suite on the design.

Statement and block coverage are the same thing.

The most popular reports are statement, path and expression coverage. Statement coverage can also be called block coverage, where a block is a sequence of statements that are executed if a single statement is executed. The code in Sample 2-10 shows an example of a statement block. The block named *acked* is executed entirely whenever the expression in the *if* statement evaluates to TRUE. So counting the execution of that block is equivalent to counting the execution of the four individual statements within that block.

Sample 2-10.
Block vs. statement execution

```
if (dtack == 1'b1) begin: acked
    as      <= 1'b0;
    data    <= 16'hZZZZ;
    bus_rq <= 1'b0;
    state   <= IDLE;
end
```

But block boundaries may not be that obvious.

Statement blocks may not be necessarily clearly delimited. In Sample 2-11, two statements blocks are found: one before (and including) the *wait* statement, and one after. The *wait* statement may have never completed and the process was waiting forever. The subsequent sequential statements may not have executed. Thus, they form a separate statement block.

Sample 2-11.
Blocks separated by a *wait* statement

```
address <= 16#FFED#;
ale     <= '1';
rw      <= '1';
wait until dtack = '1';
read_data := data;
ale     <= '0';
```

Statement Coverage

Did you execute
all the statements?
Statement or block coverage measures how much of the total lines of code were executed by the verification suite. A graphical user interface usually lets the user browse the source code and quickly identify the statements that were not executed. Figure 2-13 shows, in a graphical fashion, a statement coverage report for a small portion of code from a model of a modem. The actual form of the report from any code coverage tool or source code browser will likely be different.

Figure 2-13.
Example of
statement
coverage

```
☑ if (parity == ODD || parity == EVEN) begin
☐     tx <= compute_parity(data, parity);
☐     #(tx_time);
  end
☑ tx <= 1'b0;
☑ #(tx_time);
☑ if (stop_bits == 2) begin
☑     tx <= 1'b0;
☑     #(tx_time);
  end
```

Why did you not
execute all state-
ments?
The example in Figure 2-13 shows that two out of the eight executable statements - or 25 percent - were not executed. To bring the statement coverage metric up to 100 percent, a desirable goal, it is necessary to understand what conditions are required to cause the execution of the uncovered statements. In this case, the parity must be set to either ODD or EVEN. Once the conditions have been determined, you must understand why they never occurred in the first place. Is it a condition that can never occur? Is it a condition that should have been verified by the existing verification suite? Or is it a condition that was forgotten?

It is normal for
some statements to
not be executed.
If it is a condition that can never occur, the code in question is effectively dead: it will never be executed. Removing that code is a definite option; it reduces clutter and increases the maintainability of the source code. However, a good defensive (and paranoid) coder often includes code that is not meant to be executed. This additional code simply monitors for conditions that should never occur and reports that an unexpected condition happened should the hypothesis prove false. This practice is very effective. Functional problems are positively identified near the source of the malfunc-

tion, without having to rely on the possibility that it produces an unexpected response at the right moment when you were looking for something else.

Sample 2-12.
Defensive programming technique

```
case (mode[1:0]) // synopsys full_case
2'b00: ...
2'b10: ...
2'b01: ...
// synopsys translate_off
// coverage off
default: $write("Case was not really full!\n");
// coverage on
// synopsys translate_on
endcase
```

Your model can tell you if things are not as assumed.

Sample 2-12 shows an example of defensive modeling in synthesizeable *case* statements. Even though there is a directive instructing the synthesis tool that the *case* statement describes all possible conditions, it is possible for an unexpected condition to occur during simulation. If that were the case, the simulation results would differ from the results produced by the hardware implementation, and that difference would go undetected until a gate-level simulation is performed, or the device failed in the system.

Do not measure coverage for code not meant to be executed.

It should be possible to identify code that was not meant to be executed and have it eliminated from the code coverage statistics. In Sample 2-12, significant comments are used to remove the defensive coding statements from being measured by our hypothetical code coverage tool. Some code coverage tools may be configured to ignore any statement found between synthesis translation on/off directives. It may be more interesting to configure a code coverage tool to ensure that code included between synthesis translate on/off directives is indeed <u>not</u> executed!

Add testbenches to execute all statements.

If the conditions that would cause the uncovered statements to be executed should have been verified, it is an indication that one or more testbenches are either not functionally correct or incomplete. If the condition was entirely forgotten, it is necessary to add to an existing testbench or create an entirely new one.

Path Coverage

There is more than one way to execute a sequence of statements.

Path coverage measures all possible ways you can execute a sequence of statements. The code in Sample 2-13 has four possible paths: the first *if* statement can either be true or false. So can the second. To verify all paths through this simple code section, it is necessary to execute it with all possible state combinations for both *if* statements: false-false, false-true, true-false, and true-true.

Sample 2-13. Example of statement and path coverage

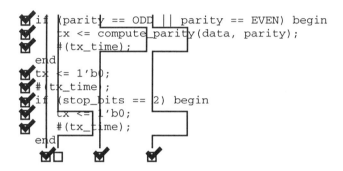

```
if (parity == ODD || parity == EVEN) begin
   tx <= compute_parity(data, parity);
   #(tx_time);
end
tx <= 1'b0;
#(tx_time);
if (stop_bits == 2) begin
   tx <= 1'b0;
   #(tx_time);
end
```

Why were some sequences not executed?

The current verification suite, although it offers 100 percent statement coverage, only offers 75 percent path coverage through this small code section. Again, it is necessary to determine the conditions that cause the uncovered path to be executed. In this case, a testcase must set the parity to neither ODD nor EVEN and the number of stop bits to two. Again, the important question one must ask is whether this is a condition that will ever happen, or if it is a conditions that was overlooked.

Limit the length of statement sequences.

The number of paths in a sequence of statements grows exponentially with the number of control-flow statements. Code coverage tools give up measuring path coverage if their number is too large in a given code sequence. To avoid this situation, keep all sequential code constructs (in Verilog: *always* and *initial* blocks, *tasks* and *functions*; in VHDL: *processes*, *procedures* and *functions*) to under 100 lines.

Reaching 100 percent path coverage is very difficult.

Expression Coverage

There may be more than one cause for a control-flow change.

If you look closely at the code in Sample 2-14, you notice that there are two mutually independent conditions that can cause the first *if* statement to branch the execution into its *then* clause: parity being set to either ODD or EVEN. Expression coverage, as shown in Sample 2-14, measures the various ways paths through the code are executed. Even if the statement coverage is at 100 percent, the expression coverage is only at 50 percent.

Sample 2-14.
Example of statement and expression coverage

```
if (parity == ODD || parity == EVEN) begin
    tx <= compute_parity(data, parity);
    #(tx_time);
end
tx <= 1'b0;
#(tx_time);
if (stop_bits == 2) begin
    tx <= 1'b0;
    #(tx_time);
end
```

Once more, it is necessary to understand why a controlling term of an expression has not been exercised. In this case, no testbench sets the parity to EVEN. Is it a condition that will never occur? Or was it another oversight?

Reaching 100 percent expression coverage is extremely difficult.

What Does 100 Percent Coverage Mean?

Completeness does not imply correctness.

The short answer is: not much. Code coverage indicates how *thoroughly* your entire verification suite exercises the source code. It does not provide an indication, in any way, about the *correctness* of the verification suite.

Results from code coverage tools should be interpreted with a grain of salt. They should be used to help identify corner cases that were not exercised by the verification suite or implementation-dependent features that were introduced during the implementation.

Code coverage lets you know if you are not done.	Code coverage is an additional indicator for the completeness of the verification job. It can help increase your confidence that the verification job is complete, but it should not be your only indicator. Code coverage indicates if the verification task is <u>not</u> complete - through low coverage numbers. A high coverage number is by no means an indication that the job is over.
Some tools can help you reach 100% coverage.	As mentioned in section titled "Testbench Generation" on page 11, there are testbench generation tools that automatically generate stimulus to exercise the uncovered code sections of a design. In my opinion, these tools are the embodiment of a misguided focus on code coverage metrics and the all-powerful and magic "100 percent". All these additional testbenches only exercise code. They do not demonstrate functional correctness. One interesting aspect of these tools occurs when they <u>cannot</u> generate a sequence of stimulus to exercise some code construct: the construct is conclusively unexecutable and should be removed (or marked as uncovered).
Code coverage tools can be used as profilers.	When developing models for simulation only, where performance is an important criteria, code coverage tools can be used for *profiling*. The aim of profiling is the opposite of code coverage. The aim of profiling is to identify the lines of codes that are executed most often. These lines of code become the primary candidates for performance optimization efforts.

VERIFICATION LANGUAGES

Verification languages can raise the level of abstraction.	As mentioned in Chapter 1, one way to increase productivity is to raise the level of abstraction used to perform a task. High-level languages, such as C or Pascal, raised the level of abstraction from assembly-level, enabling software engineers to become more productive. Similarly, computer languages specifically designed for verification are able to raise the level of abstraction compared to general-purpose simulation languages.
VHDL and Verilog are simulation languages, not verification languages.	Verilog was designed with a focus on describing low-level hardware structures. It does not provide support for high-level data structures or object-oriented features. VHDL was designed for very large design teams. It strongly encapsulates all information and communicates strictly through well-defined interfaces. Very often, these limitations get in the way of an efficient implementation of a verification strategy. Neither integrates easily with C models. This

creates an opportunity for verification languages designed to overcome the shortcomings of Verilog and VHDL. However, using verification language requires additional training and tool costs. Given the importance of the verification task, these additional costs are likely very good investments.

Proprietary verification languages exist.

At the time I wrote this book, I knew of only three proprietary verification languages: *e/Specman* from Verisity, *VERA* from Synopsys, and *Rave* from Chronology. I hope that a public-domain language will evolve (possibly from the proprietary languages) and be standardized by the IEEE. As users, we benefit most from the largest possible support for languages from as many vendors as possible. That is why VHDL and Verilog are the most popular simulation languages in the hardware design industry today. Proprietary languages constrain you to a single vendor.

You must learn the basics of verification before learning verification languages.

In this book, I use VHDL and Verilog as the implementation medium for the verification infrastructure and testbenches. Most parts could be more efficiently implemented using a verification language. But verification languages are only alternative implementation mediums. They facilitate small portions of the stimulus generation and output verification process. You still need to plan your verification, design its strategy and architecture, design the stimulus, determine the expected response, and compare the actual output. These are concepts that can be learned and applied using VHDL or Verilog. This book is primarily about these concepts and how to implement them in a language you already know (hence the decision to use both VHDL and Verilog). You need to concentrate on learning basic verification skills before learning a new language to implement them.

REVISION CONTROL

Are we all looking at the same thing?

One of the major difficulties in verification is to ensure that what is being verified is actually what will be implemented. When you compile a Verilog source file, what is the guarantee that the design engineer will use that exact same file when synthesizing the design?

When the same person verifies and then synthesizes the design, this problem is reduced to that person using proper file management discipline. However, as I hope to have demonstrated in Chapter 1, having the same person perform both tasks is not a reliable functional verification process. It is more likely that separate individuals perform the verification and synthesis tasks.

Files must be centrally managed.

In very small and closely knit groups, it may be possible to have everyone work from a single directory, or to have the design files distributed across a small number of individual directories. Everyone agrees where each other's files are, then each is left to his or her own device. This situation is very common and very dangerous: how can you tell if the designer has changed a source file and maybe introduced a functional bug since you've last verified it?

It must be easy to get at all the files, from a single location.

This methodology is not scalable either. It quickly breaks down once the team grows to more than two or three individuals. And it does not work at all when the team is distributed across different physical or geographical areas. The verification engineer is often the first person to face the non-scalability challenge of this environment. Each designer is content working independently in his or her own directories. Individual designs, when properly partitioned, rarely need to refer to some other design in another designer's working directory. As the verification engineer, your first task is to integrate all the pieces into a functional entity. That's where the difficulties of pulling bits and pieces from heterogeneous working environments scattered across multiple file servers become apparent.

The Software Engineering Experience

HDL models are software projects!

Software engineering has about 20 years of lead time dealing with the issues of managing a large number of source files, authored by many different individuals, verified by others, and compiled into a final product. Make no mistake: managing a HDL-based hardware design project is very similar to managing a software project.

Free and commercial tools are available.

To help manage files, software engineers use source control management systems. Some are available, free of charge, either bundled with all UNIX operating systems (SCCS), or distributed by the GNU project (RCS, CVS) and available in source form at:

```
ftp://prep.ai.mit.edu/pub/gnu
```

Commercial systems, some very sophisticated, are also available.

All source files are centrally managed.

Figure 2-14 shows how source files are managed using a source control management system. All accesses and changes to source files are mediated by the management system. Individual authors and users interact solely through the management system, not by directly accessing files in working directories.

Figure 2-14.
Data-flow in a
source control
system

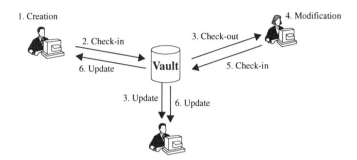

The history of a file is maintained.

Source code management systems maintain not only the latest version of a file, but also keep a complete history of each file as separate *versions*. Thus, it is possible to recover older versions of files, or to determine what changed from one version to another. It is a good idea to frequently *check-in* file versions. You do not have to rely on a back-up system if you ever accidentally delete a file. Sometimes, a series of modifications you have been working on for the last couple of hours is making things worse, not better. You can easily roll back the state of a file to a previous version known to work.

The team owns all the files.

When using a source management system, files are no longer owned by individuals. Designers may be nominally responsible for various sections of a design, but anyone - with the proper permissions - can make any change to any file. This lets a verification engineer fix bugs found in RTL code without having to rely on the designer, busy trying to get timing closure on another portion of the design, to fix them. The source management system mediates changes to files either through exclusive locks, or by merging concurrent modifications.

Configuration Management

Each user works from a *view* of the file system.

Each engineer working on a project managed with a source control system has a private *view* of all the source files (or a subset thereof) used in the project. Figure 2-15 shows how two users may have two different views of the source files in the management system. Views need not be always composed of the latest versions of all the files. In fact, for a verification engineer, that would be a hindrance. Files checked-in on a regular basis by their author may include syntax errors, be simple placeholders for future work, or be totally broken. It would be very frustrating if the model you were trying to verify kept changing faster than you could identify problems with it.

Figure 2-15.
User views of managed source files

design.v 1.53
cpuif.v 1.28
atmif.v 1.38

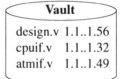

Vault

design.v 1.1..1.56
cpuif.v 1.1..1.32
atmif.v 1.1..1.49

design.v 1.41
cpuif.v 1.17
atmif.v 1.38

Configurations are created by tagging a set of versions.

All source management systems use the concept of symbolic tags that can be attached to specific versions of files. You may then refer to particular versions of files, or set of files, using the symbolic name, without knowing the exact version number they refer to. In Figure 2-15, the user on the left could be working with the versions that were tagged as "ready to simulate" by the author. The user on the right, the system verification engineer, could be working with the versions that were tagged as "golden" by the ASIC verification engineer.

Configuration management translates to tag management.

Managing releases becomes a problem of managing tags, which can be a complex task. Table 2-1 shows a list of tags that could be used in a project to identify the various versions of a file as it progresses through the design process. Some tags, such as the "*Version_M.N*" tag, never move once applied to a specific version. Others, such as the "*Submit*" tag, move to newer versions as the development of the design progresses. Before moving a tag, it may be a good idea to leave a trace of the previous position of a tag. One possible mechanism for doing so is to append the date to the tag name. For example, the old "*Submit*" version gets tagged with the new tag

"Submit_000302" on March 2nd, 2000 and the "Submit" tag is moved to the latest version.

Tag Name	**Description**
Submit	Ready to submit to functional verification. Author has verified syntax correctness and basic level of functionality.
Bronze	Passes a basic set of functional testcases. Release is sufficiently functional for integration.
Silver	Passes all functional testcases.
Gold	Passes all functional testcases and meets coding coverage guidelines (requires additional corner-case testcases).
To_Synthesis	Ready to submit to synthesis. Usually matches "Silver" or "Gold".
To_Layout	Ready to submit to layout. Usually matches "Gold".
Version_M.N	Version that was manufactured. Matches corresponding "To_Layout" release. Future versions of the same chip will move tags beyond this point.
ON_YYMMDD	Some meaningful release on the specified date.

Table 2-1. Example tags for release management

Working with Releases

Views can become out of date as new versions of files are checked into the source management system database and tags are moved forward.

Releases are specific configurations. The author of the RTL for a portion of the design would likely always work with the latest version of the files he or she is actively working on, checking them in frequently (typically at relevant points of code development throughout the day and at the end of each day). Once the source code is syntactically correct and its functionality satisfies the designer (by using a few *ad hoc* testbenches), the corresponding version of the files are tagged as ready for verification.

Users must update
their view to the
appropriate
release.

You, as the verification engineer, must be constantly on the look-out for updates to your view. When working on a particularly difficult testbench, you may spend several days without updating your view to the latest version ready to be verified. That way, you maintain a consistent view of the design under test and limit changes to the testbenches - which you make. Once the actual verification and debugging of the design starts, you probably want to refresh your view to the latest "ready to verify" release of the design before running a testbench.

You can be notified of new
releases.

An interesting feature of some source management systems is the ability to issue email notification whenever a significant event occurs. For example, such a system could send email to all verification engineers whenever the tag identifying the release that is ready for verification is moved. Optionally, the email could contain a copy of the descriptions of the changes that were made to the source files. Upon receiving such an email, you could make an informed decision about whether to update your view immediately.

ISSUE TRACKING

There be bugs
there!

The job of any verification engineer is to find bugs. Under normal conditions, you should expect to find functional irregularities. You should be <u>really</u> worried if no problems are being found. Their occurrence is normal and do not reflect the abilities of the hardware designers. Even the most experienced software designers write code that includes bugs, even in the simplest and shortest routines. Now that we've established that bugs <u>will</u> be found, how will you deal with them?

Bugs must be
fixed.

Once a problem has been identified, it <u>must</u> be resolved. All design teams have informal systems to track issues and ensure their resolutions. However, the quality and scalability of these informal systems leaves a lot to be desired.

What Is an Issue?

Is it worth worrying about?

Before we discuss the various ways issues can be tracked, we must first consider what is an issue worth tracking. The answer depends highly on the tracking system used. The cost of tracking the issue should not be greater than the cost of the issue itself. However, do you want the tracking system to dictate what kind of issues are tracked? Or, do you want to decide on what constitutes a trackable issue, then implement a suitable tracking system? The latter position is the one that serves the ultimate goal better: making sure that the design is functionally correct.

An issue is <u>anything</u> that can affect the functionality of the design:

1. Bugs found during the execution of a testbench are clearly issues worth tracking.

2. Ambiguities or incompleteness in the specification document should also be tracked issues. However, typographical errors definitely do not fit in this category.

3. Architectural decisions and trade-offs are also issues.

4. Errors found at all stages of the design, in the design itself or in the verification environment should be tracked as well.

5. If someone thinks about a new relevant testcase, it should be filed as an issue.

When in doubt, track it.

It is not possible to come up with an exhaustive list of issues worth tracking. Whenever an issue comes up, the only criteria that determines whether it should be tracked, should be its effect on the correctness of the final design. If a bad design can be manufactured when that issue goes unresolved, it <u>must</u> be tracked. Of course, all issues are not created equal. Some have a direct impact on the functionality of the design, others have minor secondary effects. Issues should be assigned a priority and be addressed in order of that priority.

You may choose not to fix an issue.

Some issues, often of lower importance, may be consciously left unresolved. The design or project team may decide that a particular problem or shortcoming is an acceptable limitation for this particular project and can be left to be resolved in the next incarnation of the product. The principal difficulty is to make sure that the decision was a conscious and rational one!

The Grapevine System

Issues can be verbally reported.

The simplest, and most pervasive issue tracking system is the *grapevine*. After identifying a problem, you walk over to the hardware designer's cubicle (assuming you are not the hardware designer as well!) and discuss the issue. Others may be pulled in the conversation or accidentally drop in as they overhear something interesting being debated. Simple issues are usually resolved on the spot. For bigger issues, everyone may agree that further discussions are be warranted, pending the input of other individuals. The priority of issues is implicitly communicated by the insistence and frequency of your reminders to the hardware designer.

It works only under specific conditions.

The grapevine system works well with small, closely knit design groups, working in close proximity. If temporary contractors or part-time engineers are on the team, or members are distributed geographically, this system breaks down as instant verbal communications are not readily available. Once issues are verbally resolved, no one has a clear responsibility for making sure that the solution will be implemented.

You are condemned to repeat past mistakes.

Also, this system does not maintain any history. Once an issue is resolved, there is no way to review the process that led to the decision. The same issue may be revisited many times if the implementation of the solution is significantly delayed. If the proposed resolution turns out to be inappropriate, the team may end up going in circles, repeatedly trying previous solutions. Without history, you are condemned to repeat it. There is no opportunity for the team to learn from its mistakes. Learning is limited to the individual, and to the extent that he or she keeps encountering similar problems.

The Post-It System

Issues can be tracked on little pieces of paper.

When teams become larger, or when communications are no longer regular and casual, the next issue tracking system that is used is the 3M Post-It™ note system. It is easy to recognize at a glance: every team member has a number of telltale yellow pieces of paper stuck around the periphery of their computer monitor.

If the paper disappears, so does the issue.

This evolutionary system only addresses the lack of ownership of the grapevine system: whoever has the yellow piece of paper is responsible for its resolution. This ownership is tenuous at best. Many issues are "resolved" when the sticky note accidentally falls on the floor and is swept away by the janitorial staff.

Issues cannot be prioritized.

With the Post-It system issues are not prioritized. One bug may be critical to another team member, but the owner of the bug may choose to resolve other issues first simply because they are simpler and because resolving them instead reduces the clutter around his computer screen faster. All notes look alike and none indicate a sense of urgency more than the others.

History will repeat itself.

And again, the Post-It system suffers from the same learning disabilities as the grapevine system. Because of the lack of history, issues are revisited many times, and problems are recreated repeatedly.

The Procedural System

Issues can be tracked at group meetings.

The next step in the normal evolution of issue tracking is the procedural system. In this system, issues are formally reported, usually through free-form documents such as e-mail messages. The outstanding issues are reviewed and resolved during team meetings.

Only the biggest issues are tracked.

Because the entire team is involved and the minutes of meetings are usually kept, this system provides an opportunity for team-wide learning. But the procedural system consumes an inordinate amount of precious meeting time. Because of the time and effort involved in tracking and resolving these issues, it is usually reserved for the most important or controversial ones. The smaller, less important - but much more numerous - issues default back to the grapevine or Post-It note systems.

Computerized System

Issues can be tracked using databases.

A revolution in issue tracking comes from using a computer-based system. In such a system, issues must be seen through to resolution: outstanding issues are repeatedly reported loud and clear. Issues can be formally assigned to individuals or list of individuals. Their resolution need only involve the required team members. The computer-based system can automatically send daily or weekly status reports to interested parties.

A history of the decision making process is maintained and archived. By recording various attempted solutions and their effectiveness, solutions are only tried once without going in circles. The resolution process of similar issues can be quickly looked-up by anyone, preventing similar mistakes from being committed repeatedly.

But it should not be easier to track them verbally on paper.

Even with its clear advantages, computer-based systems are often unsuccessful. The main obstacle is their lack of comparative ease-of-use. Remember: the grapevine and Post-It systems are readily available at all times. Given the schedule pressure engineers work under and the amount of work that needs to be done, if you had the choice to report a relatively simple problem, which process would you use:

1. Walk over to the person who has to solve the problem and verbally report it.

2. Describe the problem on a Post-It note, then give it to that same person (and if that person is not there, stick it in the middle of his or her computer screen).

3. Enter a description of the problem in the issue tracking database and never leave your workstation?

It should not take longer to submit an issue than to fix it.

You would probably use the one that requires the least amount of time and effort. If you want your team to successfully use a computer-based issue tracking system, then select one that causes the smallest disruption in their normal work flow. Choose one that is a simple or transparent extension of their normal behavior and tools they already use.

I was involved in a project where the issue tracking system used a proprietary X-based graphical interface. It took about 15 seconds to bring up the entire interface on your screen. You were then faced with a series of required menu selections to identify the precise division, project, system, sub-system, device, and functional aspect of the problem, followed by several other dialog boxes to describe the actual issue. Entering the simplest issue took <u>at least</u> three to four minutes. And the system could not be accessed when working from home on dial-up lines. You can guess how successful that system was....

Email-based systems have the greatest acceptance.

The systems that have the most success invariably use an email-based interface, usually coupled with a Web-based interface for administrative tasks and reporting. Everyone on your team uses email. It is probably already the preferred mechanism for discussing issues when members are distributed geographically or work in different time zones. Having a system that simply captures these emails, categorizes them and keeps track of the status and resolution of individual issues (usually through a minimum set of required fields in the email body or header), is an effective way of implementing a computer-based issue tracking system.

METRICS

Metrics are essential management tools.

Managers love metrics and measurements. They have little time to personally assess the progress and status of a project. They must rely on numbers that (more or less) reflect the current situation.

Metrics are best observed over time to see trends.

Metrics are most often used in a static fashion: what are the values today? How close are they to the values that indicate that the project is complete? The odometer reports a static value: how far have you travelled. However, metrics provide the most valuable information when observed over time. Not only do you know where you are, but you can know how fast you are going, and what direction you are heading (is it getting better or worse?).

Historical data should be used to create a baseline.

When compared with historical data, metrics can paint a picture of your learning abilities. Unless you know how well (or how poorly) you did last time, how can you tell if you are becoming better at your job? It is important to create a baseline from historical data to determine your productivity level. In an industry where the manufacturing capability doubles every 18 months, you cannot afford to maintain a constant level of productivity.

Metrics can help assess the verification effort.

There are several metrics that can help assess the status, progress and productivity of functional verification. One has already been introduced: code coverage.

Code-Related Metrics

Code coverage may not be relevant.

Code coverage measures how thoroughly the verification suite exercises the source code being verified. That metric should climb steadily toward 100 percent over time. From project to project, it should climb faster, and closer. However, code coverage is not a suitable metric for all verification projects. It is an effective metric for the the smallest design unit that is individually specified (such as an FPGA, a reusable component, or an ASIC). But it is ineffective when verifying designs composed of sub-designs that have been independently verified. The objective of that verification is to confirm that the sub-designs are interfaced and cooperate properly, not to verify their individual features. It is unlikely (and unnecessary) to execute all the statements.

The number of lines of code can measure implementation efficiency.

The total **number of lines of code** that is necessary to implement a verification suite can be an effective measure of the effort required in implementing it. This metric can be used to compare the productivity offered by verification languages. If they can reduce the number of lines of code that need to be written, they should reduce the effort required to implement the verification.

Lines-of-code ratio can measure complexity.

The **ratio of lines of code** between the design being verified and the verification suite may measure the complexity of the design. Historical data on that ratio could help predict the verification effort for a new design by predicting its estimated complexity.

Code change rate should trend toward zero.

If you are using a source control system, you can measure the **source code changes** over time. At the beginning of a project, code changes at a very fast rate as new functionality is added and initial versions are augmented. At the beginning of the verification phase, many changes in the code are required by bug fixes. As the verification progresses, the rate of changes should decrease as there are fewer and fewer bugs to be found and fixed. Figure 2-16 shows a plot of the expected code change rate over the life of a project. From this metric, you are able to determine if the code is becoming stable, or identify the most unstable sections of a design.

Figure 2-16.
Ideal code change rate metric over time

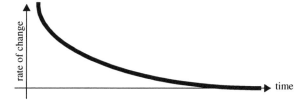

Quality-Related Metrics

Quality is subjective, but it can be measured indirectly.

Quality-related metrics are probably more directly related with the functional verification than other productivity metrics. Quality is a subjective value, yet, it is possible to find metrics that correlate with the level of quality in a design. This is much like the number of customer complaints or the number of repeat customers can be used to judge the quality of retail services. All quality-related metrics in hardware design concern themselves with measuring bugs.

A simple metric is the number of known issues.

The easiest metric is the **number of known outstanding issues**. The number could be weighed to count issues differently according to their severity. When using a computer-based issue tracking system, this metric, as well as trends and rates, can be easily generated. Are issues accumulating (indicating a growing quality problem)? Or, are they decreasing and nearing zero?

Code will be worn out eventually.

If you are dealing with a reusable or long-lived design, it is useful to measure the **number of bugs found during its service life**. These are bugs that were not originally found by the verification suite. If the number of bugs starts to increase dramatically compared to historical findings, it is an indication that the design has outlived its useful life. It has been modified and adapted too many times and needs to be re-designed from scratch. Throughout the normal life cycle of a reusable design, the number of outstanding issues exhibits a behavior as shown in Figure 2-17.

Figure 2-17. Number of outstanding issues throughout the life cycle of a design

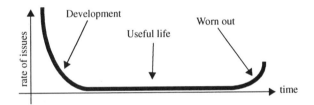

Interpreting Metrics

Whatever gets measured gets done.

Because managers rely heavily on metrics to measure performance - and ultimately assign reward and blame, there is a tendency for any organization to align its behavior with the metrics. That is why you must be extremely careful to select metrics that faithfully represent the situation and are correlated with the effect you are trying to measure or improve. If you measure the number of bugs found and fixed, you quickly see an increase in the number of bugs found and fixed. But do you see an increase in the quality of the code being verified? Were bugs simply not previously reported? Are designers more sloppy when writing their code since they'll be rewarded only when and if a bug is found and fixed?

Make sure metrics are correlated with the effect you want to measure.

Figure 2-18 shows a list of files names and current version numbers maintained by two different designers. Which designer is more productive? Do the large version numbers from the designer on the left indicate someone who writes code with many bugs that had to be fixed?

Figure 2-18.
Using version
numbers as a
metric

alu_e.vhd	1.15	cpuif_e.vhd	1.2
alu_rtl.vhd	1.234	cpuif_rtl.vhd	1.4
decoder_e.vhd	1.12	regfile_e.vhd	1.1
decoder_rtl.vhf	1.155	regfile_rtl.vhf	1.7
dpath_e.vhd	1.7	addr_dec_e.vhd	1.3
dpath_rtl.vhd	1.176	addr_dec_rtl.vhd	1.6

On the other hand, Figure 2-19 shows a plot of the code change rate for each designer. What is your assessment of the code quality from designer on the left? It seems to me that the designer on the right is not making proper use the revision control system...

Figure 2-19.
Using code
change rate as
a metric

 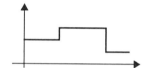

SUMMARY

In this chapter, I described different kinds of tools that can be used in a verification process. Each is used to verify a different aspect of the design and identify different types of errors. Linting tools find problems created by using questionable constructs. Simulators exercise your design to find functional errors. Waveform viewers display your simulation results in a familiar graphical presentation. Code coverage analysis helps you determine the thoroughness of your testbenches. Revision control and issue tracking systems help manage your design data. Metrics produced by these tools allow management to keep informed on the progress of a project and to measure productivity gains.

CHAPTER 3   THE VERIFICATION PLAN

In this chapter, I describe the verification plan as a specification for the verification effort. It is used to define what is first-time success, how a design is verified, and which testbenches are written.

The design project that sits before you will propel your company to new levels of market share and profitability. A few system architects have designed and specified a system that should meet performance and cost goals. Several design leaders, using the system specification, have been working on writing detailed functional specification documents for each of the ASICs and FPGAs that are required to build this new product. Teams of hot-shot hardware designers are being assembled to implement each ASIC or FPGA. Using the detailed specification documents for each device, they are coming up with a detailed implementation schedule. So far, it appears that the project will meet its production deadline.

You are in charge of the verification for this design. Not only must this product be on-time, but it must be functionally correct. Your company's reputation depends on it. You have been asked by the project manager to produce a detailed schedule for the verification and define your staffing requirements. How can you determine either?

THE ROLE OF THE VERIFICATION PLAN

Traditionally, verification is an ad-hoc process.

In a traditional verification process, your decision would be simple. In fact, your own position would not exist. Verification would be left to each hardware designer to do as they wish. It would be performed as time allows. And everybody's fingers would be crossed hoping that system integration would be smooth and that the board designs would not need too many barnacles. Many devices would be implemented in FPGAs, trading additional per-unit costs for flexibility in fixing problems later found during system integration.

Tools exist to help determine when you are done.

The tools described in the previous chapter will help during your verification effort. Code coverage, bug find rate, and code change rates are metrics that indicate how much progress you have made toward your goal. But they are like stock market indices or batting averages: they provide a snapshot of the current situation and, if recorded over time, show trends and progression. However, they cannot be used to predict the future.[1]

Specifying the Verification

You need a tool to determine when you will be done.

Today's question is about producing a schedule. You must determine, as reliably as possible, when the verification will be completed to the required degree of confidence. Unless you have a detailed specification of the work that needs to be accomplished, you cannot determine how many persons you need, nor how long it is going to take, or even if you are doing work that needs to be done. That's what the verification plan is about.

Start from the design specification.

Before you can decide on a plan of attack for the verification, a complete specification document for the design to be verified must exist. And it must exist in written form. "Folklore" specifications that describe the design as *the same thing as we did before, but at twice the clock rate and with these additional features* are insufficient. The specification document is the common source for the verification and implementation efforts. It is the golden reference and the rule of law. Later, when discrepancies are found between

---

1. Although many financial and sports analysts make a good living predicting an essentially random process or explaining, after the fact, why everybody was wrong.

the response expected by the testbench and the one produced by the design under verification, the specification document arbitrates and decides which one has the correct answer.

The verification plan is the specification document for the verification effort.

Today's million-gate ASIC designs cannot proceed without a detailed specification document being written first. With the verification effort being 100 to 200 percent of the RTL design effort, why should it proceed without a specification document of its own? The verification plan is the specification document for the verification effort.

Defining First-Time Success

If, and only if, it is in the plan, will it be verified.

The verification plan provides a forum for the entire design team to define what *first-time success* is. It is a mechanism that ensures all essential features are appropriately verified. If you want first-time success, you must identify which features must be exercised under which conditions and what the expected response should be. The verification plan documents which features are a priority and which ones are optional. In the face of schedule pressure, the decision to drop features from the first-time success requirements becomes a conscious one. The alternative is to live with whatever happens to work when the decision to ship the design cuts off the verification effort like a guillotine. Some of the features, essential for market acceptance, might fall in the basket.

From the verification plan, a detailed schedule can be created.

The verification plan creates a line-in-the-sand that cannot be crossed without endangering the success of the project in the market place. Once the plan is written, you know how many testbenches must be written, how complex they need to be, and how they depend on each other. You can define a detailed verification schedule, and allocate testbenches to resources, parallelizing verification as much as possible. Once all the testbenches verifying the must-have features are written, and the RTL passes those testcases, the design can be shipped. Not before.

The team owns the verification plan.

It is important for everyone involved with the design project to realize that they have a stake in the verification plan. The responsibility of an RTL designer is not to design RTL. That's only his job. His responsibility is to produce a working design. The entire design team must contribute to the verification plan, to make sure that it is complete and correct.

This process is not revolutionary.

The process used to write a verification plan is not new. It has been used for decades by NASA, the FAA and aerospace companies to ensure that the ultra-reliable systems they were implementing met their specifications. This process has been used for software as well as for hardware designs.

LEVELS OF VERIFICATION

Verification can be performed at various levels of granularity.

The first question, when planning the verification, is to determine the level of granularity for the verification effort. A design is potentially composed of several levels. Some have a physical partition, such as printed circuit boards, FPGAs, and ASICs. Others have a logical partition, such as synthesized units, reusable components, or sub-systems.

As illustrated in Figure 3-1, each level of verification is best suited for a particular application and objective. The nature of design with reusable components shifts where stand-alone unit-level verification ends and system-level verification starts within the physical hierarchy, compared with a more traditional design process. Design with reuse does not diminish the need for verification. The unit-to-system boundary is a logical one instead of a physical one.

Figure 3-1.
Application of different levels of verification

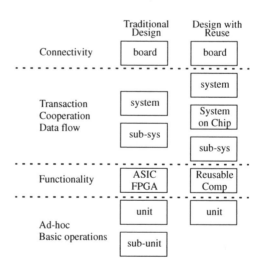

Deciding between levels of granularity involves trade-offs.

Smaller partitions are easier to verify because they offer greater controllability and observability. It is easier to setup interesting conditions and state combinations and to observe if the response is as expected. With larger partitions, the integration of the smaller partitions contained within it is implicitly verified at the cost of lower controllability and observability.

Verifying at a given level of granularity requires stable interfaces.

Because the verification requires a significant implementation effort, any partition being verified must have relatively stable interfaces and intended functionality. If the interfaces keep changing, or functionality keeps being moved from one partition to another, the testbenches will constantly need to be changed with little progress being made. Once you've decided on specific partitions to be verified, their interface and overall functionality must be specified early on and remain as stable as possible. Ideally, each verified partition should have its own specification document.

Unit-Level Verification

Implementation determines the content of this partition.

Design units are logical partitions. They are created to facilitate the implementation or the synthesis process. They vary from the relatively small (e.g. FIFOs and state machines) to the complex (e.g. PCI slave interface and DSP datapaths). Their interfaces and functionality tend to vary a lot over time, as implementation details highlight shortcomings in the initial design. They usually do not have an independent specification document to verify against either.

Use ad-hoc verification for design units.

Because these design units are usually in a constant state of flux, they are better left to an *ad-hoc* verification process. The designer himself verifies the basic operation of the unit. The objective of this verification is to ensure that there are no syntax errors in the RTL code, and that basic functionality is operational. It is not to create a regressionable test suite and obtain high code coverage.

They are too numerous to verify formally.

The high number of design units in any project makes a verification process implemented at that level too time-consuming. Each would require a custom verification environment, as described in Chapters 5 and 6. The precious verification resources would spend an inordinate amount of time creating stimulus generators and response monitors for a myriad of ever-changing interfaces. Writing a lot of simple testbenches is just as much work, if not more, as writing a few complex ones. And verification at the ASIC- or FPGA-level would still be required to verify the integration of these design units.

Unit-level verification may be required in large devices.

In today's very large and complex ASIC and FPGAs, it may not be possible to obtain the necessary functional coverage when verifying from the ASIC or FPGA partition. For the highly sensitive and complex functional blocks inside them, it may be necessary to perform unit-level verification to have sufficient levels of controllability and observability. Ideally, each functional unit verified at the unit level should have its own specification document.

Architect the design to facilitate unit-level verification.

If your design is so complex that you have to perform some unit-level verification, it should be designed to make that unit-level verification as relevant and complete as possible. Partition the design so the features to be verified are completely contained within a unit and can be verified on a stand-alone basis. Once verified, these features can be assumed to work during the verification of the higher levels. If the features to be verified at the unit level require interaction with other units, they have to be re-verified at a higher-level where the features are fully contained, to ensure that the integration correctly implements them.

Reusable Components Verification

Reusable designs are independent of any particular use.

Reusable components are designed to an independent specification. They are intended to be used as-is and unchanged in many different designs. Their reusability can be limited to a single product, the entire product family, or they could be applicable to any product requiring their functionality. They must be designed - and thus verified - independent of any one usage.

Verification components can be reused as well.

Reusable components are usually designed using standardized interfaces. These interfaces can be designed to standard on-chip buses, or industry-standard external physical interfaces. The verification components used to stimulate and monitor these interfaces can be themselves reused across the various verification environments used to verify different reusable components. The verification effort can be leveraged across multiple components, thus minimizing the overall investment in verification. Chapter 6 will detail how to architect a testbench to promote the creation and use of reusable verification components.

Reusable components need a regression test suite.	Reusable components are used in many designs. When they are modified, either to fix problems that were found, or to enhance their functionality, you must make sure that they remain backward-compatible. This is accomplished by implementing a *regression test suite* that verifies the compliance of the component after any modification. Checking the equivalence of the new version with the previous version using formal verification would not really work unless the modifications were not functional. Adding functionality or fixing problems, by definition, make the new version of the design not equivalent to the previous one.
They need thorough functional coverage.	Components will not be reused if potential users do not have confidence that they perform better and more reliably than one they could design themselves. This confidence can be obtained only by demonstrating the correctness and robustness of the components through a thorough, well documented verification process.

ASIC and FPGA Verification

The physical partition is an ideal verification level.	ASICs and FPGAs are physical partitions. They form a natural partition for the verification because their interfaces and functionality do not change very much after the initial specification of the system and the completion of their specification documents.
They may have to be treated as systems.	The ever increasing densities offered by the semiconductor technology enables ever increasing integration of complex functionality into a single device. To manage this complexity from a design and verification standpoint, they are often designed as a collection of independently designed and verified components, usually reusable but not necessarily so. In that case, the ASIC is called a *System-on-a-Chip* (SoC) and its verification resembles a system-level verification process, as described in the next section. The bulk of the functional verification is performed using unit-level verification.
FPGAs now require an ASIC-like verification process.	Traditionally, FPGAs were able to survive an ad-hoc or even the complete absence of a verification process. Their ease of programmability, often without additional component costs, allowed their functionality to be modified up to the last minute. But today's million-gate FPGAs, even with only 50 percent effective usage, can implement functions that are too complex to verify and debug during integration. Their functionality must be verified from the RTL code, before synthesis and implementation.

System-Level Verification

A system need not follow physical boundaries.

Everybody's definition of a *system* is different. In this book, a system is a <u>logical</u> partition composed of independently verified components. A system could thus be composed of a few reusable components and cover a subset of an SoC ASIC. A system could also be composed of several ASICs physically located on separate printed circuit boards, as illustrated in Figure 3-2.

Figure 3-2.
Logical
system
partition

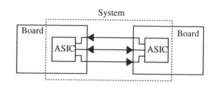

The verification focuses on interaction.

The system-level verification focuses on the interactions between the individual components instead of the functionality implemented in each one. The latter is better verified at the component-level verification. The system verification engineer has to rely on the individual components being functionally correct.

The testcase defines the system.

Since systems are logical partitions, they can be composed of any number of components, regardless of their physical location. Which system to use and verify depends on the testcases that are determined to be interesting and significant. To minimize the simulation overhead, it is preferable to use the smallest possible system necessary to execute the specified testcase. However, the number of possible systems being very large, a set of "standard" systems should be defined. The same system is used for many testcases even if, in some cases, some of the included components are not required.

Board-Level Verification

Board-level models are generated from the board design tool.

The primary objective of board-level verification is to confirm that the connectivity captured by the board design tool is correct. An entire board can also be used as a system to verify its functionality. Unlike a logical system model, the model for the board design must be automatically generated by the board capture tool. When verifying the board design, or any other physical partition, you must make sure that what is being verified is what will be manufactured. There must be a direct link between the captured design and what is simulated. Automatic generation of the board-level model from the capture tool provides that link. A logical system model has no such restriction: it can be manually generated for the system of interest.

Many components on a board do not fit in a digital simulation environment.

The main difficulty with board-level models is obtaining suitable models for all the components. That is where third-party sources and hardware modelers are useful (see "Third-Party Models" on page 36). Also, generating a model out of a board design tool involves introducing approximations. For example, how do you represent capacitors in a digital simulation environment? Analog devices, connectors, opto-couplers, and other components used in board-level designs do not translate easily in a digital simulation environment either.

Board-level parasitics can be modeled.

The generated model may include models for board-level parasitics that may affect the functional correctness of the board. As the speed of signals in a board increases, transmission line effects are becoming important. ASICs can no longer be designed without consideration of their eventual use on a circuit board.

How about a connectivity formal verification tool?

I've always been of the opinion that using functional simulation to verify connectivity is a poor verification strategy. I would like to see a tool that, given a formal description of the intended connectivity (e.g. this pin of this ASIC is connected to that pin of that CPLD) and a netlist automatically extracted from a board design tool, would compare the two and identify any discrepancies. This would accomplish the task in a static fashion, without requiring stimulus or verifying the response. Errors could not go unnoticed simply because they were not exercised during the simulation. I am convinced that the time to independently capture a formal specification of the board connectivity would take less time than writing a single testbench for it. Existing formal verification tools could probably be easily extended to include this functionality.

VERIFICATION STRATEGIES

Decide on a black- or white-box approach for various levels of granularity.

Given the functionality that needs to be verified, you must decide on a strategy for carrying out the verification. You must decide on the level of granularity where verification will be accomplished. You must also decide on the types of testcases that will be used for each level of granularity. Testcases can be either white-box or black-box, depending on the visibility and knowledge you have of the internal implementation of each unit under verification (see

"Black-Box Verification" on page 12 and "White-Box Verification" on page 13).

Decide on the level of abstraction where the tescases will be specified.	You also need to decide the level of abstraction where the bulk of the verification will be performed. The higher levels of abstraction usually apply to coarser granularity of design under verification. With higher levels of abstraction, your have less detailed control over the timing and coordination of the stimulus and response, but it is much easier to generate a lot of stimulus and verify long responses. If detail controls are required to perform certain testcases, it may be necessary to work at a lower level of abstraction.
A processor interface could be verified at the cycle or device driver level.	For example, verifying a processor interface can be accomplished at the individual read and write cycle levels. But that requires each testcase to have an intimate knowledge of the memory-mapped registers and how to program them. That same interface could be driven at the device driver level. The testcase would have access to a set of high-level procedural calls to perform complete operations. Each operation is composed of many individual read and write cycles to specific memory-mapped registers, but the testcase is removed from these implementation details.

Verifying the Response

Plan how you will check the response.	Deciding how to apply the stimulus is relatively easy. You are under complete control of its timing and content. It is verifying the response that is difficult. You must plan how you will determine the expected response, then how to verify that the design provided the response you expected. The section titled "Predicting the Output" on page 211 suggests several techniques for implementing output verification.
Some responses are difficult to verify in the simulation.	Throughout this book, implementing self-checking testbenches is recommended (see "Verifying the Output" on page 172). But, it can sometimes be difficult for a testbench to verify a response that can be immediately recognized as right or wrong by a human. For example, verifying a graphic engine involves checking the output picture for expected content. A self-checking simulation would be very good at verifying individual pixels in the picture. But a human would be more efficient in recognizing a filled red circle. The verification strategy must find a way to automate these type of testcases.

| Detect errors as early as possible. | It may be more efficient to have the simulation produce a set of outputs that can be later compared against a set of reference outputs. The result of a simulation can be further processed outside of the simulator to determine success or failure. However, it is more efficient to detect problems as early as possible. When the response is checked within the simulation, the error is identified while the model is near the state that produced the error. It is easier to identify and fix the error. |

Random Verification

Random verification still provides valid stimulus.	A strategy often used for system-level verifications is random verification. Random verification does not mean that you randomly apply zeroes and ones to every input signals in the design. This would not represent an accurate usage of the design and would not accomplish anything. With random verification, the inputs are subjected to valid individual operations, such as a read cycle or an Ethernet packet. It is the *sequence* of these operations and the content of the data transferred that is random.
They will create conditions you did not think of.	Random simulations are used to create conditions that you have not thought about when writing your verification plan. They create unexpected conditions or hit corner cases. They also reduce the bias introduced by the verification engineer when coding the testbenches. Instead of creating input sequences that are easy to code, they create more realistic stimulus. Random stimulus can also be used to create background activity on other interfaces, while you focus on creating well-specified testcases on one or two interfaces.
They are complex to specify and code.	Because of their nature, random simulations are complex to specify. The sequence of data and operations applied to the design must be a fair representation of the operating conditions the design will be under. The specification for a random simulation must include distributions and ranges of data and operations. More complicated is the prediction of the expected output. Since you do not know what stimulus will be generated, how can you know what the response will be? The strategy used for random testcases must address how an invalid response is detected.
Random simulations are usually used at the system level.	Because of the complexity of implementation, random verification is usually applied at coarser levels of granularity. Many design units and smaller partitions can thus take advantage of a single random simulation environment. A possible strategy, encouraged by the

verification language *e* from *Verisity*, is to start with a random verification. Based on the verification plan, source code, and functional coverage metrics, the random simulation is tuned and constrained into individual directed testcases.

FROM SPECIFICATION TO FEATURES

Identify features.

The first step in writing a verification plan is to identify the features that will be verified. From the specification document, you enumerate all the features that are described and thus must be verified. Other team members, especially the system architects and RTL designers, contribute additional features to be verified. These additional features may not have been obvious in the specification to someone unfamiliar with the purpose or characteristics of the design. Other features may become significant once a particular implementation is chosen.

Label each feature.

Features should be labelled and have a short description. The feature should be described in terms what needs to be verified, not how it is to be implemented. Each feature should be cross-referenced to the section or paragraph describing it in details in the specification document. Ideally, the specification document should also contain a cross-reference to the feature list in the verification plan. A subset of the feature list for a Universal Asynchronous Receiver Transmitter (UART) is shown in Sample 3-1.

Sample 3-1.
Some of the
features of a
UART design

1. The Clear-To-Send (CTS) pin must be asserted when the UART can accept a new word to be transmitted via the CPU interface. See section 4.2 of the UART specification document.

2. The CTS bit in the status register must reflect the level of the CTS pin. See table A.2

3. The Data Ready (DTR) pin must be asserted when there is a received word ready to be read by the CPU interface. See section 4.1 of the UART specification document.

4. The DTR bit in the status register must reflect the level of the DTR pin. See table A.2

5. Data bits are serially sent and received with the least significant bit first. See section 5.2 of the UART specification document.

Specify features for the proper level of verification.

When enumerating features, be careful to include them in the verification plan for the proper verification level. Some features are better verified at the component (unit, reusable, ASIC) level, while others must be verified at the system level.

Component-Level Features

They are fully contained within the unit being verified.

A component can be a unit, a reusable component, or an entire ASIC. Component-level features are fully contained within the component being verified. They do not involve system-level interaction with other components. Their correctness can be determined without depending on a subsequent verification of the integration of the component into a high-level system. Examples of component-level features include the ones listed in Sample 3-1.

The bulk of the features will be component-level features. These features are assumed to be functional when the component is used in a system-level verification.

System-Level Features

Minimize system-level features.

A system can be a subset of an ASIC, a few ASICs from different boards, an entire board design, or the complete product. Because of the large size and long runtime of system-level simulations, it is necessary to minimize the features verified at this level. Whenever something is identified as a system-level feature, question whether it can be verified as a component-level feature instead. For example, in the design illustrated in Figure 3-3, the *MX* ASIC can select between the data from ASICs *ID0* or *ID1* under software control. Is the switching feature a system-level feature? The answer is *no*. The switching feature is entirely contained within the *MX* ASIC and is thus a component-level feature.

Figure 3-3.
Example of a
system
structure

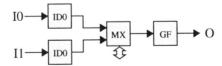

System-level features include connectivity, flow control and inter-operability.

System-level features are usually limited to connectivity, flow-control, and inter-operability. For example, the connectivity from the input ports to the output port would be a system-level feature. In verifying the connectivity, it is necessary to switch the input from the *ID0* stream to the *ID1* stream. But the switching is not the primary objective of the verification and would be assumed to work.

Another system testcase would be verifying that full input FIFOs in the *MX* ASIC creates back-pressure through the *ID0* and *ID1* ASICs and stop the flow of data until the congestion clears.

Error Types to Look For

Assume design tools do not introduce functional errors.

When listing features to be verified, there is an implicit assumption about the errors that are likely to occur and should be found. Functional verification must focus on finding functional errors in the design. It is not the responsibility of the functional verification to make sure that the design tools are bug-free. Functional simulation ensures that the design was implemented as specified without interpretation errors or problems introduced by the designers. For example, running all functional testbenches on the gate-level netlist only verifies that the synthesis tool works properly. Formal verification and static timing analysis are better tools to accomplish this task.

Likely errors are different based on the capture tool used.

The types of errors that can be made are different when using different capture tools. When schematic capture tools are used, connectivity errors, such as reversed bit orders in a bus, or misconnected individual bits within a bus, are very common. In a RTL coding and logic synthesis environment, this type of error is not likely to occur: if a bus is properly connected, either all the bits work, or none do. Linting tools can detect some connectivity problems such as multiple drivers on a wire or an output that goes nowhere and would be a better strategy for identifying these types of problems.

Look for functional errors.

Common errors in a synthesis-based design flow include wrong polarities, protocol violations or incorrect computations. The type of stimulus that proved useful in the days of schematic capture, such as walking ones and zeroes may not be as useful in a RTL design verification. A pair of patterns of alternating ones and zeroes, for example "0xAAAA" followed by "0x5555", is usually sufficient. Using *signatures* in the data stream is another efficient technique to detect functional error. A signature can be as simple as a sequential number to help detect missing or repeating data items. A signature can also encode either the source or the expected destination of a data item. For example, the data associated with an address in a write cycle could contain a portion of the address and an identification of the bus master issuing the cycle. The section titled "Packet Processors" on page 215 details how to use signatures to verify a class of designs.

FROM FEATURES TO TESTCASES

Prioritize

Prioritize the features.

Not all features are created equal. Once they are enumerated, they must be prioritized. Some features are *must-have* for the design to properly function or to meet the demands of the market. This is the stage that defines first-time success. These features must operate properly for the design to be shipped. The completion of the verification of these features gates the successful completion of the project and the testbenches verifying these features are often on the critical path. The *must-have* features need to be thoroughly verified for all possible configuration and usage options.

Less important features receive less attention.

The *should-have* features are not primordial for the commercial success of the design. They may simply offer expansion capabilities or differentiation from the competition. The main objective is to verify their basic functionality for correct operation. If time and resources allow, more detailed verification of these features may be accomplished. The verification of these features may be cancelled if schedule pressure forces the reallocation of resources to the verification of more important features.

Some features are verified only as time allows.

The *nice-to-have* features are purely optional. They are verified only as time allows, usually in a primitive fashion. The reality of today's design schedule almost guarantees that they'll never be verified!

Make an informed decision when cutting back on the verification effort.

The prioritization of the features to be verified lets a project manager make informed decisions when schedule pressures make it necessary to eliminate some planned activities. The verification effort can be trimmed starting with features that were pre-determined to be less important. If a greater impact of the project completion date is required and *must-have* features are dropped from the verification, the decision will be a conscious one as these priorities were clearly identified as critical to the initial marketing objectives. Cutting the verification effort of *must-have* features requires a conscious re-evaluation of the marketing objectives for the project.

Group into Testcases

Groups features with similar verification requirements.

Features naturally fall into groups. Some features require similar configuration, granularity, or verification strategy to perform their verification. To maximize productivity, these features should be grouped together and assigned to the same verification engineer. For example, all features related to the CPU interface should be grouped together. As another example, verifying the baud rate, number of data bits and parity generation of a UART falls within the same group. Each group of feature verification forms a *testcase*.

Testcases can be specified from code coverage metrics.

Some testcases will not come directly from features extracted from the specification. For example, testcases created to increase code coverage are not usually attached to a particular set of features. These testcases should be clearly identified as such.

Cross-reference into the feature list.

Each testcase should be labelled and given a short description of its objective. Its description should contain a list of the features verified in this testcase. The feature list should also be annotated with cross-references to the testcases where a particular feature is being verified. If a feature does not have a cross-reference to a testcase, it is not being verified.

Define dependencies.

The description of a testcase should also contain a list of the features assumed to be operational and functionally correct. From these dependencies, you can determine the order in which the testcase must be written, and identify any parallelism opportunities in the testbench development effort.

Specify the testcase stimulus.

The sequence and characteristics of the stimulus for the testcase must also be described. For example, describe the various operations or bus cycles that must be performed. For random testcases, specify the range and distribution of the input data and types of operations.

Specify the acceptance criteria.

More than just the expected response, the testcase specification must state how the response will be determined as valid. This includes expected values, timing, and protocol. For example, the output of a packet processor could be determined as correct solely on the basis of the destination address matching the output port where it appeared. Or, a more stringent requirement could be specified, such as packets from different sources showing up in the proper order and interleaved with a proper distribution.

Specify what
should not happen.

One of the more explicit ways of describing acceptance criteria is to state exactly which errors to look for. For example, making sure that a packet comes out with a correct CRC value. Another example is to describe events that are mutually exclusive, such as the assertion of the *full* and *empty* flags in a FIFO. Being explicit about what errors to look for lets a verification engineer, who is not intimately familiar with the design, implement a highly reliable testbench.

Inject errors to
make sure they are
detected.

Never trust a testbench that does not produce error messages. Every testcase should include some error injection mechanism to make sure that errors are detected and reported by the testbench. The absence of an error message would be a failure condition for that testcase. For example, a testcase verifying the parity generation in a UART should purposefully misconfigure the parity in the UART to make sure that the testbench detects a wrong parity.

Design for Verification

Hard-to-verify fea-
tures will be iden-
tified.

At this stage of the verification planning, hard-to-verify features will be identified. They can be difficult to verify because the testbench lacks suitable controllability or observability of the features. An example would be the verification that an embedded 64-bit counter properly rolls over and that the processing algorithm works properly across the roll-over point. The difficulty may be because of a poor choice in verification granularity. In that case, a smaller partition containing the hard-to-verify features should be used. The difficulty may also be due to the choice of implementation architecture or an artifact of the design itself. If a smaller partition cannot be used, or would not ease the verification of these features, a grey-or white-box approach must be taken.

Modify the design
to aid verification.

The advantage of planning the verification up front is that you can still influence the implementation of the design. If some features prove to be to difficult to verify given the current architecture and feature set of the design, have the design modified to include additional features to aid in their verification. Hardware design engineers will no doubt complain about adding functionality that is not really needed by the design. However, if the alternative is to create a design you cannot verify, what choice do they have? These features have always proven to be useful during lab integration of sample parts.

Provide counter pre-load functions.

If the design contains long counters or other state elements with large numbers of states, make sure they can be pre-loaded to an arbitrary value via a memory-mapped register. Ideally, their current value should be available for read back through the same register set. In the previous example, a series of 8 bytes in the address space of the design could be allocated to pre-loading and reading back the value of the 64-bit counter.

Provide datapath by-pass paths.

The correct implementation of long data paths can also be difficult to verify if you do not have detailed control over all the operands. For example, speech synthesizers are simple digital signal processing designs with a datapath that shapes random noise[2]. You have complete control over the coefficients applied to the data samples to form specific sounds. However, you do not have control over one critical element: the primary input data value. That's a random number. To properly verify the operation of this datapath, you need control over its initial input value.

Figure 3-4.
Verifiable
datapath for a
speech
synthesizer

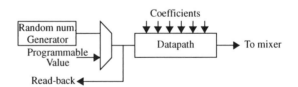

As shown in Figure 3-4, the design should include a mechanism to use a programmable constant input values instead of a random number as input to the datapath. Conversely, you should also be able to read back the output of the random number generator to ensure that it is indeed producing random numbers. Pop quiz: why is the read-back point located after the multiplexor that selects between the normal operation using the random number generator

---

2. It is used to produce consonant sounds, such as the *sh* sound. It is then mixed with a shaped base frequency used to produce vowel sounds, such as the *a* sound, to hopefully create intelligible speech.

and the programmable static value, and not at the output of the random number generator?[3]

Provide sample points.

If observability is the problem but not controllability, adding sample points readable through memory-mapped registers can help ease the verification of some features. If the address space allocated to the design is at a premium, these sample points could be multiplexed into a single address location, using a second address to select which point is currently being sampled.

Provide error injection mechanism.

If the design includes error detection mechanisms, you may want to have provisions in an upstream design to inject errors. The decision to include error injection should be carefully considered. If it is for hardware verification only, it may not be properly documented for the software engineers. This feature may be accidentally turned on when a device driver writes a value that was thought to be inoffensive.

FROM TESTCASES TO TESTBENCHES

Testcases naturally fall into groups.

Just like features, testcases naturally fall into groups. They require a similar configuration of the design, use the same abstraction level for the stimulus and response, generate similar stimulus, determine the validity of the response using a similar strategy, or verify closely-related features. For example, the testcase verifying that a UART properly transmits data can be grouped with the testcase that verifies its configuration controls. Both need similar stimulus (a variety of data words to transmit), and both verify the correctness of the output in a similar fashion (is the data value identical, with no parity error).

Group testcases into testbenches.

Each group of testcases is then divided into testbenches. A popular division, the one used in this book, is one testcase per testbench. The minimization of Verilog compilation time, or the time spent back-annotating a large gate-level netlist with a correspondingly large Standard Delay File (SDF) may dictate that a minimum num-

---

3. You want to verify that, when the datapath is put into normal operation mode, the multiplexor is functionally correct and the input value is indeed coming from the random number generator.

ber of testbenches be created by grouping several testcases into a single testbench.

Cross-reference testbenches with testcases.

Each testbench should be labelled and uniquely identified. This identifier should be used as the filename where the top-level HDL code for the testbench is implemented. For each testbench, enumerate the list of testcases it implements. Then cross-reference each testbench into the testcase list. The description of a testcase should contain the name of the testbench where it is implemented. If a testbench is not identified, a testcase has not yet been implemented.

Allocate each group to an engineer.

Regardless of the division of testcases into testbenches, allocate each group of testcases to a verification engineer. Testcases in the same group have similar implementation requirements. They can build on the implementation of previous testcases in the group. The first testbench takes the longest to write. But as the engineer responsible for the testcase group gains experience and debugs his or her verification infrastructure, a lot can be reused, often through cut-and-paste, in subsequent testbenches. The name of the individual to whom a testbench has been assigned should be recorded in the verification plan. That person is responsible for implementing the testbench according to its specification.

Verifying Testbenches

How do you verify that testbenches implement the verification plan?

The purpose of the verification effort and writing testbenches is to verify that a design meets its specification. If the verification plan is the specification for the verification effort, how do you verify that the testbenches implement their specification? How can you prevent a significant portion of a testcase from being skipped because of human error? Testbenches often include temporary code structures to by-pass large sections to speed up the debugging of a critical section. How can you make sure that they are taken out, returning the testbench to implementing the entire set of testcases it is supposed to contain?

Using a model that is known to be broken would be too cumbersome.

One possible strategy is to use a known broken model. But you would have to have a broken model for each feature that is verified in a verification plan. Each would have to be maintained so that the unbroken features match the latest version of the design under verification. You could be tempted to introduce these breaks directly into the model of the design itself with some control mechanism to select which one to turn on or off. But that would increase the risk of introducing a fault in the design that would be manufactured. Using a broken model to verify testbenches is not practical because of the complexity of managing a large number of controlled and known failure modes inside a design.

Verify testbenches through peer reviews. As described in "The Human Factor" on page 5, one way to verify a transformation performed by a human (in this case, writing a testbench from a specification), is to provide redundancy. Once completed, testbenches should be reviewed by other verification engineers to ensure that they implement the specification of the testcases they contain. For more details, refer to section "Code Reviews" on page 28. The simulation output log should also be reviewed to ensure that the execution of the testbench follows the specification as well. To that effect, the testbench should produce regular notice messages. It should state what stimulus is about to be generated, and what error or response is being checked. The output log should ultimately contain, in a bullet form, the specification of the testcases that have been executed.

SUMMARY

In this chapter, I have outlined a process for writing a verification plan. This plan is the specification document for the verification process. It is used to define first-time success and to influence the design specification to include features that will facilitate verification. This chapter also defined how a system-level testcase is different from a unit-level testcase. Once the verification plan is written, a schedule for the verification can be created: you know how many testbenches must be written and how complex they are.

## CHAPTER 4

BEHAVIORAL HARDWARE DESCRIPTION LANGUAGES

A proper verification engineer must break the "RTL mindset" that most hardware engineers, out of necessity, have grown into. To efficiently accomplish the verification task, you must be well-versed in behavioral (i.e. non-synthesizeable and highly algorithmic) descriptions. To reliably and correctly use the behavioral constructs of VHDL or Verilog, it is necessary to understand the side effects of the simulation algorithm and the limitations of the language - and ways to circumvent them. This understanding was not required to successfully write RTL models.

BEHAVIORAL VERSUS RTL THINKING

In this section, I illustrate the difference between the approach to writing a RTL model compared with writing a behavioral model.

Many guide-
lines help code
RTL models.

All experienced hardware design engineers are very comfortable with writing synthesizeable models. They conform to a well-defined subset of the VHDL or Verilog languages and follow one of a few coding styles. Numerous RTL coding guidelines have been published.[1] They help designers obtain efficient implementations: low area, high speed, or low power. Guidelines, such as the ones shown in Sample 4-1, can help a novice designer avoid undesirable hardware components, such as latches, internal buses, or tristate buffers. More importantly, guidelines can also help maintain an identical behavior between the synthesizeable model and the gate-level implementation, such as the ones shown in Sample 4-2.

Sample 4-1.
RTL coding guidelines to avoid undesirable hardware structures

1. To avoid latches, set all outputs of combinatorial blocks to default values at the beginning of the block.

2. To avoid internal buses, do not assign *reg*s from two separate *always* blocks (Verilog only).

3. To avoid tristate buffers, do not assign the value 'Z' (VHDL) or 1'bz (Verilog).

Sample 4-2.
RTL coding guidelines to maintain simulation behavior

1. All inputs must be listed in the sensitivity list of a combinatorial block.

2. The clock and asynchronous reset must be in the sensitivity list of a sequential block.

3. Use a non-blocking assignment when assigning to a *reg* intended to be inferred as a flip-flop (Verilog only).

The adherence to the synthesizeable subset and proper coding guidelines can be easily verified using a linting tool (more details are in the section titled "Linting Tools" on page 22). After several months of experience, the subset becomes very natural to hardware designers. It matches their mental model of a hardware design: state machines, operators, multiplexers, decoders, latches, clocks, etc..

Do not use RTL-like code when writing test-benches.

The synthesizeable subset is adequate for describing the implementation of a particular design. I often claim that VHDL and Verilog are both equally poor at this task. The subset was dictated by the synthesis technology, not by someone with a warped sense of humor playing a practical joke on the entire industry. It was designed to describe hardware structures and logical transformations between registers, matching the capability of logic synthesis technology. However, this subset becomes quickly insufficient

---

1. The IEEE will soon publish (or may have already published) a standard set of guidelines for RTL coding. For Verilog, see *"IEEE P1364.1 Standard for Verilog Register Transfer Level Synthesis"* prepared by the Verilog Synthesis Interoperability Working Group of the Design Automation Standards Committee. For VHDL, see *"IEEE P1076.6 Standard for VHDL Register Transfer Level Synthesis"* prepared by the VHDL Synthesis Interoperability Working Group of the Design Automation Standards Committee.

when writing testbenches that were never intended to be implemented in hardware. Both languages have a rich set of constructs and statements. If you have an "RTL mindset" when writing testbenches and limit yourself to using a coding style designed to describe relatively low-level hardware structures, you will not take full advantage of the language's power. The verification task will be needlessly tedious and complicated.

Contrasting the Approaches

The example below shows a simple handshaking protocol. Your task is to write some VHDL or Verilog code that implements the simple handshaking protocol shown in Figure 4-1. It detects that an acknowledge signal (ACK) is asserted (high) after a requesting signal (REQ) is asserted (high). Once the acknowledge is detected, the requesting signal is deasserted, and it then waits for the acknowledge signal to be deasserted.

Figure 4-1.
State diagram for handshaking protocol

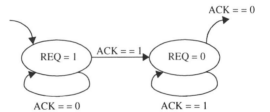

RTL-Thinking Example. A hardware designer, with an RTL mindset, will immediately implement the state machine shown in Figure 4-1. The corresponding VHDL code is shown in Sample 4-3. This relatively simple behavior required 28 lines of code and two processes to describe, and two additional states in a potentially more complex state machine.

Focus on behavior, not implementation.

Behavioral-Thinking Example. A verification engineer, with a behavioral mindset, will instead focus on the *behavior* of the protocol, not its implementation as a state machine. The corresponding code is shown in Sample 4-4. The functionality can be described behaviorally using only four statements.

Behavioral models are faster to write.

Modeling this simple protocol using behavioral constructs should require less than 10 percent of the time required to model it using synthesizeable constructs. Not only is there less code to write (14

Sample 4-3.
Synthesize-
able VHDL
code for sim-
ple handshak-
ing protocol

```
type STATE_TYP is (..., MAKE_REQ, RELEASE, ...);
signal STATE, NEXT_STATE: STATE_TYP;
...
COMB: process (STATE, ACK)
begin
    NEXT_STATE <= STATE;
    case STATE is
    ...
    when MAKE_REQ =>
        REQ <= '1';
        if ACK = '1' then
            NEXT_STATE <= RELEASE;
        end if;
    when RELEASE =>
        REQ <= '0';
        if ACK = '0' then
            NEXT_STATE <= ...;
        end if;
    ...
    end case;
end process COMB;

SEQ: process (CLK)
begin
    if CLK'event and CLK = '1' then
        if RESET = '1' then
            STATE <= ...;
        else
            STATE <= NEXT_STATE;
        end if;
    end if;
end process SEQ;
```

Sample 4-4.
Behavioral
VHDL code
for simple
handshaking
protocol

```
process
begin
    ...
    REQ <= '1';
    wait until ACK = '1';
    REQ <= '0';
    wait until ACK = '0';
    ...
end process;
```

percent), but it is also simpler, requiring less effort to ensure that it is correct.

Behavioral mod-
els simulate faster.

Another benefit of behavioral modeling is the increase in simulation performance. Assuming that there is a long delay between a

change in the request and the corresponding acknowledgement, the simulation of the synthesizeable model would still execute the *SEQ* process at every transition of the clock (because that process is sensitive to the clock signal). The process containing the behavioral description would wait for the proper condition of the acknowledge signal, resuming execution only when the protocol is satisfied. If the acknowledge signal replies after a 10-clock-cycle delay, this represents a reduction of process execution from 40 in the synthesizeable version to two in the behavioral one, or a 1900 percent increase in simulation performance.

YOU GOTTA HAVE STYLE!

The synthesizeable subset puts several constraints on the coding style you may use. Even with these restrictions, many less experienced hardware designers manage to write RTL code that is difficult to understand and maintain. There are no such restrictions with behavioral modeling. With this complete and thorough freedom, it is not surprising that even experienced designers produce testbench code that is unmaintainable, fragile, and not portable.

A Question of Discipline

Write maintainable, robust code.

There are no laws against writing bad code. If you do, the consequences do not involve personal fines or prison terms. However, the consequences do involve a real economic cost to your employer. Your code <u>will</u> need to be modified: either to fix a functional error, to extend its functionality, or to adapt it to a new design. When (not *if*) your code needs to be modified, it will take the person in charge of making that modification more time than would otherwise have been required had the code been written properly the first time. Under extreme conditions, your code may even have to be re-written entirely.[2]

My first job after graduating from university was to design and implement a portion of a logic synthesis tool using the C language. In those days, I had been writing code in various languages for over eight years and I measured my performance as a software engineer

---

2. Do not think "*It won't be my problem*". You may very well be that person and you may not be able to understand your own code weeks later.

by the cleverness of my implementations of algorithms. I felt really proud of myself when I was able to craft a complex computation into a "poetic" one-liner. C is the ultimate software craftsman language!

Invest time now, save support time later.

I soon came to realize the error of my ways. During the eight previous years, I always wrote "disposable" code: the programs were either short-lived (school assignments or personal projects), or they had a narrow audience (utilities for university professors or a learning aid for a particular class). Never had I written a program that would live for several years and be used by dozens of persons, each with their own sophisticated needs and attempting to use my program in ways I had never intended or even thought of. As I found myself having to fix many problems reported by users, I had difficulties understanding my own code written only weeks before. I quickly learned that time invested in writing better code would be saved many times over in subsequent support efforts.

Optimize the Right Thing

You should <u>always</u> strive for maintainability. Maintainability is important even when writing synthesizeable code. Before optimizing some aspect of your code, make sure it really needs improvement. If your code meets all of its constraints, it does not need to be optimized. Maintainability is the most important aspect of any code you write because understanding and supporting code is the most expensive activity.

Saving lines actually costs money.

There is no economic reason to reduce the number of lines of code. Unless, of course, it also improves the maintainability. Saving one line of code, with an average of 50 characters per line, saves only 50 bytes on the storage medium. With 10Gb hard drives costing less than \$300 today, this represents a savings of $\$28 \times 10^{-9}$. The time saved in typing, assuming an extremely slow typing speed of one character per second and a loaded labor rate for an engineer at \$100,000 a year, amounts to \$0.69. However, if saving that line reduces the understandability of the code where it will require an additional 5 minutes to figure out its operation, the additional cost incurred amounts to \$4.17. The total <u>loss</u> from reducing code by 1 line equals \$3.48. And that is for a single line, and a single instance of maintenance.

Optimizing perfor-
mance costs
money.

Similar costs are incurred when optimizing code for performance. These optimizations usually reduce maintainability and must be done only when absolutely required. If the code meets its constraints as is, do not optimize it. That principle applies to synthesizeable code as well. The example in Sample 4-5 was taken from the user's manual of a verification language. It is a synthesizeable description of a 2-bit round-robin arbiter.

Sample 4-5.
Synthesize-
able code for
2-bit round-
robin arbiter

```verilog
module rrarb(request, grant, reset, clk);
input  [1:0] request;
output [1:0] grant;
input        reset;
input        clk;
wire         winner;
reg          last_winner;
reg    [1:0] grant;
wire   [1:0] next_grant;

assign next_grant[0] =
   ~reset & (request[0] &
             (~request[1] | last_winner));

assign next_grant[1] =
   ~reset & (request[1] &
             (~request[0] | ~last_winner));

assign winner =
   ~reset & ~next_grant[0] &
   (last_winner | next_grant[1]);

always @ (posedge clk)
begin
   last_winner = winner;
   grant = next_grant;
end
endmodule
```

RTL code can be
too close to sche-
matic capture.

Some aspects of maintainable code were used in Sample 4-5: identifiers are meaningful and the code is properly indented. However, the continuous assignment statements implementing the combinatorial decoding demonstrate that the author was thinking in terms of boolean equations, maybe even working from a schematic design, not in terms of functionality of the design.

This approach complicates making even the simplest change to this design. If you need further convincing that this design is difficult to understand, try to figure out what happens to the content of the *last_winner* register when there are no requests. Another potential problem are the race conditions created by using the blocking assignments in the *always* block (for more details, see "Read/Write Race Conditions" on page 141).

Specify function first, optimize implementation second - and only if needed.

The code shown in Sample 4-6 implements the same function, but it is described with respect to its functionality, not its gate-level implementation. It is easier to understand and to determine whether it is functionally correct by simple inspection.

If you need further convincing that this design is easier to understand, try to figure out what happens to the content of the *last_winner* register when there are no requests. Was it easier to determine with this model or the one in Sample 4-5?

Sample 4-6. Synthesizeable code for 2-bit round-robin arbiter

```
module rrarb(request, grant, reset, clk);
input  [1:0] request;
output [1:0] grant;
input        reset;
input        clk;

reg [1:0] grant;
reg last_winner;
always @ (posedge clk)
begin
    if (reset) begin
        grant       <= 2'b00;
        last_winner <= 0;
    end
    else begin
        grant <= 2'b00;
        if (request != 2'b00) begin: find_winner
            reg winner;
            case (request)
            2'b01: winner = 0;
            2'b10: winner = 1;
            2'b11:
                if (last_winner == 1'b0) winner = 1;
                else                     winner = 0;
            endcase
            grant[winner] <= 1'b1;
            last_winner   <= winner;
        end
    end
end
endmodule
```

It is also easier to modify, for example, should the request and grant signals be asserted low instead of high. The synthesized results should be close to that of the previous model. It should not be a concern until it is demonstrated that the results do not meet area, timing or power constraints. Your primary concern should be maintainability, unless shown otherwise.

Good Comments Improve Maintainability

If reducing the number of lines of code actually increases the overall cost of a design, the same argument applies to comments. One could argue that reducing the number of lines of code can yield a better program, since there are fewer statements to understand. However, the primary purpose of comments is explicitly to improve maintainability of code. No one can argue that reducing their number can lead to better code.

You can write bad comments.

However, just as there is bad code, there are bad comments. Obsolete or outdated comments are worse than no comments at all since they create confusion. Comments that are cryptic or assume some particular knowledge may not be very useful either. One of the most common mistakes in commenting code, illustrated in Sample 4-7, is to describe in written language what the code actually does.

Sample 4-7.
Poor comment

```
-- Increment addr
addr := addr + 1;
```

Unless you are trying to learn the language used to implement the model, this comment is self-evident and redundant. It does not add any information. Any reader familiar with the language would have understood the functionality of the statement. Comments should describe the intent and purpose of the code, as illustrated in Sample 4-8. It is information that is not readily available to someone unfamiliar with the design.

Sample 4-8.
Proper comments

```
-- In burst mode, the bytes are written in
-- consecutive addresses. Need to access the
-- next address to verify that the next byte
-- was properly saved.
addr := addr + 1;
```

Assume an inexpe-
rienced audience.

When commenting code, you should assume that your audience is composed of junior engineers who are familiar with the language, but not with the design. Ideally, it should be possible to strip a file of all of its source code and still understand its functionality based on the comments alone.

STRUCTURE OF BEHAVIORAL CODE

This section describes techniques to structure and encapsulate behavioral code for maximum maintainability. Encapsulation can be used to hide implementation details and package reusable code elements.

RTL models
require a well-
defined structure
strategy.

Structuring code is the process of allocating portions of the func-tionality to different modules or entities. These modules or entities are then connected together to provide the complete functionality of the design. There are many guidelines covering the structure of syn-thesizeable code. That structure has a direct impact on the ease of meeting timing requirements. The structure of a synthesizeable model is dictated by the limitations of the synthesis tools, often with little regard to the functionality.

Testbenches are
structured accord-
ing to functional
needs.

A testbench implemented using behavioral Verilog or VHDL code does not face similar tool restrictions. You are free to structure your code any way you like. For maintainability reasons, behavioral code is structured according to functionality or need. If a function is particularly complex, it is easier to break it up in smaller, easier to understand subfunctions. Or, if a function is required more than once, it is easier to code and verify it separately. Then you can use it as many times as necessary with little additional efforts. Table 4-1 shows the equivalent constructs available in each language to help structure code appropriately.

Table 4-1.
Available
constructs for
structuring
code

VHDL	Verilog
Entity and architecture	Module
Function	Function
Procedure	Task
Package and package body	Module

Encapsulation Hides Implementation Details

Encapsulation is an application of the structuring principle. The idea behind encapsulation is to hide implementation details and decouple the usage of a function from its implementation. That way, the implementation can be modified or optimized without affecting the users, as long as the interface is not modified.

Keep declarations as local as possible

The simplest encapsulation technique is to keep declarations as local as possible. This technique avoids accidental interactions with another portion of the code where the declaration is also visible. A common problem in Verilog is illustrated in Sample 4-9: two *always* blocks contain a *for-loop* statement using the register i as an iterator. However, the declaration of i is global to both blocks. They will interfere with each other's execution and produce unexpected results.

Sample 4-9.
Improper
encapsulation
of local
objects

```
integer i;

always
begin
    for (i = 0; i < 32; i = i + 1) begin
        ...
    end
end

always
begin
    for (i = 15; i >= 0; i = i - 1) begin
        ...
    end
end
```

In Verilog, put local declarations in named blocks.

In Verilog, you can declare registers local to a *begin/end* block if the block is named. A proper way of encapsulating the declarations of the iterators so they do not affect the module-level environment is to declare them locally in each *always* block, as shown in Sample 4-10. Properly encapsulated, these local variables cannot be accidentally accessible by other *always* or *initial* blocks and create unexpected behavior.

Other locations where you can declare local registers in Verilog include *tasks* and *functions*, after the declaration of their arguments.

Sample 4-10.
Proper encap-
sulation of
local objects

```
always
begin: block_1
    integer i;
    for (i = 0; i < 32; i = i + 1) begin
        ...
    end
end

always
begin: block_2
    integer i;
    for (i = 15; i >= 0; i = i - 1) begin
        ...
    end
end
```

An example can be found in Sample 4-11. In VHDL, declarations can be located before any *begin* keyword.

Sample 4-11.
Local declara-
tions in tasks
and functions

```
task send;
    input [7:0] data;

    reg         parity;
begin
    ...
end
endtask

function [31:0] average;
    input [31:0] val1;
    input [31:0] val2;

    reg  [32:0] sum;
begin
    sum = val1 + val2;
    average = sum / 2;
end
endfunction
```

Encapsulating Useful Subprograms

Some functions and procedures are useful across an entire project or between many testbenches. One possibility would be to replicate them wherever they are needed. This obviously increases the required maintenance effort. It also duplicates information that was already captured. VHDL has *packages* to encapsulate any declaration used in more than one entity or architecture. Verilog has no such direct features, but it provides other mechanisms that can serve a similar purpose.

Example: error reporting routines.

One example of procedures that is used by many testbenches are the error reporting routines. To have a consistent error reporting format (which can be easily parsed later to check the result of a regression), a set of standard routines are used to issue messages during simulation. In VHDL, they are implemented as procedures in a package. In Verilog, they are implemented as tasks, with two packaging alternatives.

Verilog packages can be simple `include files.

The simplest packaging technique is to put the tasks in a file to be included via a compiler directive within the module where they are used. This implementation, shown in Sample 4-13 and used in Sample 4-12, has two drawbacks:

1. First, the package cannot be compiled on its own since the tasks are not contained within a module.

2. Second, since the tasks are compiled within each module where it is included, it is not possible to include global variables, such as an error counter.

Sample 4-12.
Using tasks packaged using `include file in Verilog

```
module testcase;

`include "syslog.vh"

initial
begin
    ...
    if (...) error("Unexpected response\n");
    ...
    terminate;
end
endmodule
```

This implementation has one clear advantage compared to the alternative implementation presented in the following paragraphs: it can be used in synthesizeable code whereas the other cannot.

Tasks can be packaged in a module and used using a hierarchical name.

The other packaging technique is to put the tasks in a module to be included in the simulation, but never instantiated within any of the modules where they are used. Instead, an absolute hierarchical name is used to access the task in this global module. This imple-

Sample 4-13.
syslog.vh:
packaging
tasks using
`include file in
Verilog

```
task warning;
    input [80*8:1] msg;
begin
    $write("WARNING at %t: %s", $time, msg);
end
endtask

task error;
    input [80*8:1] msg;
begin
    $write("-ERROR- at %t: %s", $time, msg);
end
endtask

task fatal;
    input [80*8:1] msg;
begin
    $write("*FATAL* at %t: %s", $time, msg);
    terminate;
end
endtask

task terminate;
begin
    $write("Simulation completed\n");
    $finish;
end
endtask
```

Sample 4-14.
Packaging
tasks using a
module in Ver-
ilog.

```
module syslog;

integer warnings;
integer errors;
initial
begin
    warnings = 0;
    errors   = 0;
end

task warning;
    input [80*8:1] msg;
begin
    $write("WARNING at %t: %s", $time, msg);
    warnings = warnings + 1;
end
endtask
...
endmodule
```

mentation, shown in Sample 4-14 and used in Sample 4-15, can be compiled on its own since the tasks are now contained within a module. It is also possible to include global variables, such as an error counter. This implementation technique is also consistent with the one used to encapsulate bus-functional models, as explained in the following section.

Sample 4-15.
Using tasks packaged using a module in Verilog

```
module testcase;

initial
begin
   . . .
   if (...) syslog.error("Unexpected response");
   . . .
   syslog.terminate;
end
endmodule
```

Encapsulating Bus-Functional Models

In Chapter 5, I describe how data applied to the design under verification via complex waveforms and protocols can be implemented using tasks or procedures. These subprograms, called *bus-functional models*, are typically used by many testbenches throughout a project. If they model a standard interface, such as a PCI bus or a Utopia interface, they can even be reused between different projects. Properly packaging these subprograms facilitates their use and distribution.

Figure 4-2 shows a block diagram of a bus-functional model. On one side, it drives and samples low-level signals according to a predefined protocol. On the other, subprograms are available to initiate a transaction with the specified data values. The latter is called a *procedural interface*.

Figure 4-2.
Block diagram of a bus-functional model

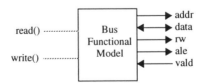

```
library ieee;
use ieee.std_logic_1164.all;
package cpu is

subtype byte is std_logic_vector(7 downto 0);

procedure write(variable wadd: in    natural;
                variable wdat: in    natural;
                signal   addr: out   byte;
                signal   data: inout byte;
                signal   rw  : out   std_logic;
                signal   ale : out   std_logic;
                signal   vald: in    std_logic);

end cpu;

library ieee;
use ieee.numeric_std.all;
package body cpu is

procedure write(variable wadd: in    natural;
                variable wdat: in    natural;
                signal   addr: out   byte;
                signal   data: inout byte;
                signal   rw  : out   std_logic;
                signal   ale : out   std_logic;
                signal   vald: in    std_logic)
is
    constant Tas: time = 10 ns;
begin
    if vald /= '0' then
        wait until vald = '0';
    end if;
    addr <= std_logic_vector(unsigned(wadd, 8));
    data <= std_logic_vector(unsigned(wdat, 8));
    rw   <= '0';
    wait for Tas;
    ale  <= '1';
    wait until vald = '1';
    ale  <= '0';
end write;

end cpu;
```

In VHDL, use pro-
cedures with *sig-
nal* arguments.

In VHDL, the bus-functional model would be implemented using a
procedure located in a package. For the procedure to be able to
drive the interface signals, they must be passed through the procedure's interface as formals of class *signal*. If the procedure had been
declared in a process, you could drive the signals directly using side

effects. It would have been possible in that context since the driver on each signal is clearly identified with the process containing the procedure declaration. Once put in a package, the signals are no longer within the scope of the procedure, nor are the drivers within the procedure attached to any process. Using *signal*-class formals lets a process pass its signal drivers to the procedure for the duration of the transaction. Sample 4-16 shows an example of properly packaged bus-functional models in VHDL.

Task arguments
are passed *by*
value only.

In Verilog, you might be tempted to implement the bus-functional model using a task where the low-level signals are passed to the tasks, similar to VHDL's procedure. However, Verilog arguments are passed *by value* when the task is called and when it returns. At no other time can a value flow into or out of a task via its interface. For example, the task shown in Sample 4-17 would never work. The assignment to the *bus_rq* variable cannot affect the outside until the task returns. The task cannot return until the *wait* statement sees that the *bus_gt* signal was asserted. But the value of *bus_gt* cannot change from the value it had when the task was called.

A simple modification to the packaging can work around the problem. Instead of passing the signals through the interface of the task, they are passed through the interface of the module implementing the package, as shown in Sample 4-18. This also simplifies calling the bus-functional model tasks as the (potentially numerous) signals need not be enumerated on the argument list for every call.

Sample 4-17.
Task arguments in Verilog are passed
by value

```
module arbiter;

task request;
    output bus_rq;
    input  bus_gt;
begin
    // The new value does not "flow" out
    bus_rq <= 1'b1;
    // And changes do not "flow" in
    wait bus_gt == 1'b1;
end
endtask
endmodule
```

<table>
<tr><td>

Sample 4-18.
Signal inter-
face on Ver-
ilog package

</td><td>

```verilog
module arbiter(bus_rq, bus_gt);
output bus_rq;
input  bus_gt;

task request;
begin
    // The new value "flows" out through the pin
    bus_rq <= 1'b1;
    // And changes "flow" in as well
    wait bus_gt == 1'b1;
end
endtask
endmodule
```

</td></tr>
</table>

DATA ABSTRACTION

Synthesizeable
models are limited
to bits and vectors.

The limitation of logic synthesis technology has forced the synthesizeable subset into dealing only with data formats that are clearly
implementable: bits, vectors of bits, and integers. Behavioral code
has no such restrictions. You are free to use any data representation
that fits your need.

Work at the same
level as the design
under verification.

You must be careful not to let an RTL mindset artificially limit your
choice, or to keep you from moving to a higher level of abstraction.
You should approach the verification problem at the same level of
granularity as the "unit of work" for the design. For an ATM cell
router, the unit of work is an entire ATM cell. For a SONET framer,
the unit of work is a SONET frame. For a video compressor, the
unit of work is either a video line or an entire picture, depending on
the granularity of the compression. The interesting conditions and
testcases are much easier to set up at that level than at the low-level
bit interface.

Abstracting data in
Verilog requires
creativity and dis-
cipline.

VHDL provides excellent support for abstracting data into high-
level representations. Verilog does not have as many features, but
with the proper technique and discipline, a lot can be accomplished.
The following sections use Verilog to illustrate how various data
abstractions can be implemented since it is the more limiting language. Their implementations in VHDL is much easier and you are
invited to consult a book on the VHDL language[3] to learn the
details.

---

3. A title is suggested in the Preface on page xix.

Real Values

This section shows how floating-point values can be used to abstract algorithmic computations. It also shows how to work around the limitations of Verilog's *real* type. The section ends with a technique to translate this abstract representation into its corresponding physical implementation.

Use *real* when verifying DSP designs.

If you are verifying any digital signal processing design, the *real* type is your friend. It is much easier to compute the expected output value of a filter using floating-point arithmetic than trying to accomplish the same thing in fixed-point representation using bit vectors, as the implemention is sure to do. Furthermore, does the latter technique offer a truly independent path to the expected answer? Or, is it simply reproducing the implementation - and thus not providing any functional verification?

For example, you have to verify a design that implements Equation 1, where a_0, a_1, a_2, b_1, and b_2 are programmable, the a and b values are between +2.0 and -2.0, and the x and y values are between +1.0 and -1.0. How would you compute the expected response (y_n) to an input sequence?

$$y_n = a_0 x_n + a_1 x_{n-1} + a_2 x_{n-2} + b_1 y_{n-1} + b_2 y_{n-2} \qquad \textbf{(EQ 1)}$$

Constants could be defined using `define symbols.

First, it is necessary to know the value of the coefficients to use. These would be determined using a digital signal processing analysis tool using a process that is well beyond the scope of this book. Usually, the coefficients are defined in the verification plan. Since the coefficients are constant for the duration of the testcase, one way to define them is to use `define symbols, as shown in Sample 4-19.

Sample 4-19. Coefficients defined using `define symbols

```
`define a0   0.500000
`define a1   1.125987
`define a2  -0.097743
`define b1  -1.009373
`define b2   0.009672
```

Defining them as parameters is better.

However, `define symbols are <u>global</u> to the compilation and violate the data encapsulation principle (you can read more about this in "Encapsulation Hides Implementation Details" on page 93). They also pollute the global name space, preventing the use of an identical symbol by someone else. A better approach is to declare them as parameters as shown in Sample 4-20. They would be local to the module.

Sample 4-20.
Coefficients
defined using
parameters

```
parameter a0 =  0.500000,
          a1 =  1.125987,
          a2 = -0.097743,
          b1 = -1.009373,
          b2 =  0.009672;
```

Implement the filter equation as a function.

A *function* could be used to compute the next output value, as shown in Sample 4-21. Because all registers are static in Verilog (i.e. they are allocated at compile time and a single copy exists in memory at all time), the internal state of the filter is kept in registers declared locally within the function. A hierarchical access is used to reset them.

But Verilog already presents one of its limitations: real values cannot be passed across interfaces. A function can return a real value, but it cannot accept a real value as one of its input arguments. Nor can tasks. Module ports cannot accept a real value either. To work around this limitation, Verilog provides a built-in system task to translate a real value to and from a 64-bit vector: *$realtobits* and *$bitstoreal,* respectively. Whenever a real value needs to be passed across an interface, it has to be translated back and forth using these system tasks.

Use a constant array in VHDL.

In VHDL, using constants defined as arrays of reals, as shown in Sample 4-22, would be the proper implementation. It has the advantage that a *for-loop* could be used to compute the equation, a solution that remains simple and efficient, independent of the number of terms in the filter. If a similar requirement is present in a Verilog environment, you could use a memory of 64-bit registers, where each memory location would contain a coefficient translated into bits.

The implementation of the function to compute the next output value in VHDL requires modifications from the Verilog implementation. In VHDL, all local subprogram variables are dynamic. They are created every time the function or procedure is called. It is not possible to maintain the state of the filter as a variable local to the

```
function real yn;
   input [63:0] xn;

   real xn_1, xn_2;
   real yn_1, yn_2;
begin
   // Compute next output value
   yn = a0 * $bitstoreal(xn) +
        a1 * xn_1 +
        a2 * xn_2 +
        b1 * yn_1 +
        b2 * yn_2;

   // Shift state of the filter
   xn_2 = xn_1; xn_1 = $bitstoreal(xn);
   yn_2 = yn_1; yn_1 = yn;
end
endfunction

initial
begin: test_procedure
   real y;

   // Reset the filter
   yn.xn_1 = 0.0; yn.xn_2 = 0.0;
   yn.yn_1 = 0.0; yn.yn_2 = 0.0;

   // Compute the response to an impulse
   y = yn($realtobits(1.0));
   repeat (99) begin
      y = yn($realtobits(0.0));
   end
end
```

```
type real_array_typ is
   array(natural range <>) of real;

constant a: real_array_typ(0 to 2) :=
   (0=>0.500000, 1=>1.125987, 2=>-0.097743);
constant b: real_array_typ(1 to 2) :=
   (1=>-1.009373, 2=>0.009672);
```

function. It is not possible to use a globally visible variable either: VHDL functions cannot have side effects. In other words, they cannot assign to any object that is not either local or an argument. However, a procedure declared in a process can have side effects.

Sample 4-23.
Procedure
computing the
next output
value

```
test_procedure: process
    variable xn_1, xn_2: real := 0.0;
    variable yn_1, yn_2: real := 0.0;

    procedure filter(xn: in  real;
                     yn: out real) is
       variable y: real;
    begin
       -- Compute next output value
       y := a(0) * xn   +
            a(1) * xn_1 +
            a(2) * xn_2 +
            b(1) * yn_1 +
            b(2) * yn_2;

       -- Shift state of the filter
       xn_2 := xn_1; xn_1 := xn;
       yn_2 := yn_1; yn_1 := y;

       -- Return next value via output argument
       yn := y;
    end filter;

    variable y: real;
begin
    -- Compute the response to an impulse
    filter(1.0, y);
    for I in 1 to 99 loop
       filter(0.0, y);
    end loop;
end process test_procedure;
```

Sample 4-23 shows the computation of the next value implemented using a procedure.

Use a conversion
function to trans-
late into a fixed-
point representa-
tion.

Eventually, it is necessary to program the coefficients, represented using real numbers, into the design under verification where a fixed-point representation is required. This is best accomplished through a conversion function, as shown in Sample 4-24. The argument is the real coefficient value and the return value is the fixed-point representation of that same value. A similar function is needed for the data values. The complementary function is required to convert the output data value from its fixed-point representation back to a real number to be compared against the expected value. The comparison could be performed using fixed-point representation, but the error message would not be as meaningful if an hexadecimal value were reported instead of a real value.

Sample 4-24.
Converting a
real value to a
fixed point
value

```
// Fixed-point number format: si.fffffffffffff
//                             || |<-----+---->|
// Sign (1=negative) ---------+|       |
// Integer portion -----------+        |
// Fractional portion ----------------+
// e.g. bit 14 is 2**0
//      bit 13 is 2**-1 (0.5)
//      bit 12 us 2**-2 (0.25)
// etc...
function [15:0] to_fixpnt;
    input [63:0] coeff;

    real c;
begin
    to_fixpnt = 16'h0000;
    c = $bitstoreal(coeff);
    if (c < 0) begin
        to_fixpnt[15] = 1'b1;
        c = -c;
    end
    to_fixpnt[14:0] = c * 'h4000;
end
endfunction
```

The testbench
becomes much
higher level.

With these utility functions in place, the testbench becomes much easier to implement and understand. Refer to Sample 4-25 for an example. The *if* statement used to compare the output value with the expected value should be modified to take into account quantization effects of performing the computation using fixed-point arithmetic compared to floating points with full precision. The comparison should accept as valid a difference within +/- 2^{-10}. The range of acceptable error should be specified in the verification plan.

Records

Records are used to represent information composed of various smaller pieces of different types. This section develops a technique for modelling records in Verilog.

Records are ideal for representing packets or frames, where control or signaling information is grouped with user information. The code Sample 4-26 shows the VHDL declaration for a record used to

Sample 4-25.
DSP testcase

```
initial
begin: test_procedure
    real xn;

    // Initialize prediction function
    yn.xn_1 = 0.0; yn.xn_2 = 0.0;
    yn.yn_1 = 0.0; yn.yn_2 = 0.0;

    // Reset the design
    ...

    // Program the coefficient
    cpu_write('A0_ADDR, to_fixpnt(a0));
    cpu_write('A1_ADDR, to_fixpnt(a1));
    cpu_write('A2_ADDR, to_fixpnt(a2));
    cpu_write('B1_ADDR, to_fixpnt(b1));
    cpu_write('B2_ADDR, to_fixpnt(b2));

    // Verify the response to an impulse input
    xn = 1.0;
    repeat (100) begin: test_one_sample
        real expect;

        data_in = to_fixpnt(xn);
        expect = yn($realtobits(xn));
        @ (posedge clk);
        if (expect !== to_real(data_out)) ...
        xn = 0.0;
    end
    $finish;
end
```

represent an ATM cell. An ATM cell is a fixed-length 53-byte packet with 48 bytes of user data.

Sample 4-26.
VHDL record
for an ATM
cell

```
type atm_payload_typ is array(0 to 47) of
    integer range 0 to 255;

type atm_cell_typ is record
    vpi    : integer range 0 to 4095;
    vci    : integer range 0 to 65535;
    pt     : bit_vector(2 downto 0);
    clp    : bit;
    hec    : bit_vector(7 downto 0);
    payload: atm_payload_typ;
end record;
```

Records can be faked in Verilog.	Verilog does not support records directly, but they can be faked. Hierarchical names can be used to access any declaration in a module. A module can emulate a record by containing only register declarations. When instantiated, the module instance emulates a record register, with each register in the module becoming a field of the record instance. The record declaration for an ATM cell can be emulated, then used in Verilog, as shown in Sample 4-27. The module containing the declaration for the record can contain instantiation of lower record module, thus creating multi-level record structures.

Sample 4-27.
Verilog record
for an ATM
cell

```
module atm_cell_typ;
  reg [11:0] vpi;
  reg [15:0] vci;
  reg [ 2:0] pt;
  reg        clp;
  reg [ 7:0] hec;
  reg [ 7:0] payload [0:47];
endmodule

module testcase;

atm_cell_typ cell();

initial
begin: test_procedure
    integer i;
    cell.vci = 0;
    ...
    for (i = 0; i < 48; i = i + 1) begin
       cell.payload[i] = 8'hFF;
    end
end
endmodule
```

The faked record is not a real object.	Although Verilog can fake records, they remain fakes. The record is not a single variable such as a register. Therefore, it cannot be assigned as a single unit or aggregate, nor used as a single unit in an expression. For example, Sample 4-28 attempts to compare the content of two cells, assign them and use them as arguments. It would produce a syntax error because the cells are instance names, not variables nor valid expressions.

Sample 4-28.
Verilog
records are not
objects

```
module testcase;

atm_cell_typ actual_cell();
atm_cell_typ expect_cell();
atm_cell_typ next_cell();

initial
begin: test_procedure
    ...
    // Verilog records cannot be compared
    if (actual_cell !== expect_cell) ...
    ...
    // Nor assigned
    expect_cell = next_cell;
    ...
    // Nor passed through interfaces
    receive_cell(actual_cell);
    ...
end
endmodule
```

Provide conver-
sion function to
and from an equiv-
alent vector.

You can work around the limitation of these fake records by using a technique similar to the one built-in for the real numbers: conversion functions between records and equivalent vectors. These functions can be located in the record definition module and called using a hierarchical name. The code Sample 4-29 shows how the *tobits* and *frombits* conversion functions can be defined and used for the ATM cell record.

Use a symbol to
predefine the size
of the equivalent
record.

One difficulty created by the workaround is knowing, as a user, how big the bit vector representation of the record is. Should the representation of the record change (e.g. a field is added), all wires and registers declared to carry the equivalent bit vector representation would need to be modified.

The best solution is to define a symbol to hide the size of the corresponding vector from the user as shown in Sample 4-30. This method presents a disadvantage: the symbol must either be declared in a file to be included using the `include directive, or the module defining the record type must be compiled before any module making use of it. The latter also requires that the `resetall directive not be used.

```
module atm_cell_typ;
reg [11:0] vpi;
reg [15:0] vci;
...
reg [ 7:0] payload [0:47];

reg [53*8:1] bits;

function [53*8:1] tobits;
    input dummy;
begin
    bits = {vpi, vci, ..., payload[47]};
    tobits = bits;
end
endfunction

task frombits;
    input [53*8:0] cell;
begin
    bits = cell;
    {vpi, vci, ..., payload[47]} = bits;
end
endtask
endmodule

module testcase;

atm_cell_typ actual_cell();
atm_cell_typ expect_cell();
atm_cell_typ next_cell();

initial
begin: test_procedure
    ...
    // Comparing Verilog records
    if (actual_cell.tobits(0) !==
        expect_cell.tobits(0)) ...
    ...
    // and assigned
    expect_cell.frombits(next_cell.tobits(0));
    ...
    // as well as passed through interfaces
    receive_cell(actual_cell.bits);
    actual_cell.frombits(actual_cell.bits);
end
endmodule
```

Sample 4-30.
Hiding the
size of the
equivalent
vector

```
module atm_cell_typ;

`define ATM_CELL_TYP [53*8:1]

reg [11:0] vpi;
reg [15:0] vci;
 . . .
reg [ 7:0] payload [0:47];

reg `ATM_CELL_TYP bits;

function [53*8:1] tobits;
   . . .
endfunction

task frombits;
   . . .
endtask
endmodule

module testcase;

atm_cell_typ actual_cell();
reg `ATM_CELL_TYP actual;

initial
begin: test_procedure
   . . .
   // Receive the next ATM cell
   receive_cell(actual);
   actual_cell.frombits(actual);
end
endmodule
```

Compiler symbols provide an alternative implementation technique.

A different approach can be used if records are not nested. A single-level record can be faked by using a vector composed of the concatenated fields, much like the equivalent bit vector of the record implemented as an instantiated module. The fields are declared and accessed using compiler symbols. The advantage of this technique is that the records are true objects, and can thus be passed through interfaces, or used in expressions. Sample 4-31 shows how a record for an ATM cell would be defined and used using this technique. Notice that, because of the linear structure of the vector, it is not possible to use a memory for the payload data. Nor is it possible to provide indexing into the payload using an expression.

Sample 4-31.
Alternative
implementa-
tion of Ver-
ilog record for
an ATM cell

In file "atm_cell_typ.vh":

```
`define ATM_CELL_TYP [53*8:1]
`define VPI      [ 12:  1]
`define VCI      [ 28: 13]
`define PT       [ 31: 29]
`define CLP      [ 32: 32]
`define HEC      [ 40: 33]
`define PAYLD_0  [ 48: 41]
`define PAYLD_1  [ 56: 49]
...
`define PAYLD_47 [424:417]
```

In file "testcase.v":

```
module testcase;

`include "atm_cell_typ.vh"

reg `ATM_CELL_TYP actual_cell;
reg `ATM_CELL_TYP expect_cell;

initial
begin: test_procedure
    ...
    // Receive the next ATM cell
    receive_cell(actual_cell);
    // Compare against expected one
    if (actual_cell != expect_cell) ...

    // Increment the VPI field
    actual_cell`VPI = actual_cell`VPI + 1;
    ...
end
endmodule
```

You cannot declare or use variant records.

Neither VHDL nor Verilog support variant records. Variant records provide different fields based on the content of another. For example, ATM cells have two flavors: UNI and NNI. Both are 53 bytes long and have a 5-byte header. They differ simply in the format of their headers. Variant records would enable representing these two different formats concurrently, using the same storage space.

Because variant records are not supported, a record structure that can represent both header formats must contain two different declarations. Variant records could also handle packets of varying lengths, such as Ethernet or IP packets. Records meant to represent variable-length packets must be declared to be large enough to handle the largest possible packet, resulting in wasted memory usage and lower performance.

VHDL records containing access types have limitations.

A variable-length record could be implemented in VHDL using an access type for the variable-length fields. However, this would put severe limitations on its usability as signals cannot be of a type containing an access type. Only variables, which are local to a single process, could make use of this variable-length record. Global shared variables could be used to bypass this limitation on using access types, but they are only available with VHDL-93. Not to mention that using global variables is also a severe violation of software engineering etiquette!

Multi-Dimensional Arrays

Single-dimensional arrays are useful data structures for representing linear information such as a fixed-length data sequences, lookup tables, or memories. Two-dimensional arrays are used for planar data such as graphics or video frames. Three-dimensional arrays are not frequently used, but they could find an application in representing data for video compression applications such as MPEG. Arrays with greater numbers of dimensions have rare applications, especially in hardware verification. This section shows how to implement multi-dimensional arrays in Verilog. It builds on the technique presented in the previous section to demonstrate how to create arrays of records.

Verilog has severe limitations on implementing arrays.

VHDL provides excellent support for arrays with any number of dimensions containing any scalar data type. Verilog, on the other hand, presents a definite challenge and using an alternate language, such as VHDL, e, or VERA, should be considered. However, if the need for a two-dimensional array is for a limited portion of the verification infrastructure, it is probably not worth investing in learning a new language and developing a new environment for the sake of a single data structure.

Verilog can easily represent some kinds of arrays.

Verilog can easily implement a one-dimensional array of bits using a multi-bit *reg*. It can implement a one-dimensional array of either vectors or integers using a memory. It could implement a two-dimensional array of bits using a memory of vectors, where the memory implements the first dimension, and the vectors in the memory implement the second one.

Multi-dimensional arrays can be mapped onto a single dimensional structure.	However, Verilog cannot directly implement a two-dimensional arrays of RGB values, where each RGB value is a record containing three 8-bit fields (red, green and blue). Again, discipline and creativity can help circumvent the limitations of the language. It is necessary to implement, in Verilog, what the compiler of a higher-level language would do to implement a multi-dimensional array.

Remember that the memory of the underlying computer simulating your testbench is linear: it goes from address 0 to hundreds of millions. A two-dimensional array has to be mapped onto this linear structure. Figure 4-3 shows how a 4x4 two-dimensional array can be mapped on a linear memory and Equation 2 shows how the location of an array element in the linear memory can be computed. You have to use a similar technique to map both dimensions of an array onto the single dimension of a Verilog multi-bit register or a memory.

Figure 4-3.
Mapping a
4x4 array to a
linear memory

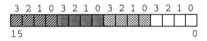

$$location = 4Y + X \qquad \text{(EQ 2)}$$

Use a Verilog memory to implement the array	If a record is going to be part of a more complex data structure, it has to be implemented using processor symbols, as shown in Sample 4-31. If you defined your RGB array using this technique, you can then use a Verilog memory to represent your two-dimensional array of RGB values: both dimensions are mapped to the memory, while each of the memory elements is an RGB record. The partial declaration for the memory is shown in Sample 4-32.

Sample 4-32.
Partial declaration for the
RGB array.

```
reg `RGB_TYP rgb_array [...];
```

Reusability of the array implementation becomes a concern.	The main problem becomes: how large should the memory be? It is not possible to know without knowing the dimensions of the array beforehand. One solution would be is to hardcode the dimension of the array being implemented, but that greatly limits the usefulness

and reusability of the data structure. What if you need two different array sizes in the same simulation? What about being able to use it in a subsequent project? You want to minimize (hopefully eliminate) the number of changes to code that has to be reused in a different context. The data structure implementation should be made independent of the array dimensions.

Module encapsulation is the preferred implementation method.

Two implementation avenues are available, identical to the alternatives available for implementing records.

1. The first one is to encapsulate the array in a module. This is the cleanest and most disciplined implementation technique.

2. The second is to use compiler symbols. However, since the array is implemented as a Verilog memory, which cannot be passed through interfaces nor treated as a single object, the latter implementation technique offers no advantages over the module encapsulation.

Implementing the array using a register instead of a memory is another alternative. However, because Verilog does not allow accessing a slice of a vector using expressions, as shown in Sample 4-33, looking up and assigning individual array elements would be very inefficient. However, this implementation could be attractive if it is necessary to pass the array often through many interfaces.

Sample 4-33.
Illegal vector slice reference in Verilog

```
function [IT_SIZE:1] lookup;
    input [31:0] x, y;

    integer index;
begin
    index = X_SIZE * y + x;
    lookup = array[(index+1) * IT_SIZE - 1 :
                    index * IT_SIZE];
end
endfunction
```

Use parameters to define the array size.

Because the array is implemented using a module, you can use parameters to define the dimensions of the array. The parameter values can be different for each instance of that module, creating different array sizes. The array module can be made more generic if a parameter is also used to define the size of the memory content. A function and a task can then be used to look-up and assign the array, respectively. Sample 4-34 shows the implementation of a generic two-dimensional array in Verilog. It would be a simple task to modify this implemention to implement a generic three-dimensional array.

Sample 4-34.
Implementation of a generic two-dimensional array

```
module ARRAY_2;

parameter X_SIZE  = 2,
          y_SIZE  = 2,
          IT_SIZE = 8;

reg [IT_SIZE:1] array [0:(X_SIZE*Y_SIZE)-1];

function [IT_SIZE:1] lookup;
   input [31:0] x;
   input [31:0] y;
begin
   lookup = array[X_SIZE*y + x];
end
endfunction

task set;
   input [     31:0] x;
   input [     31:0] y;
   input [IT_SIZE:1] it;
begin
   array[X_SIZE*y + x] = it;
end
endtask

endmodule
```

The implementation of the two-dimensional array should also contain a function and a task to convert the memory to and from a vector if the array needs to be passed through interfaces. The code in Sample 4-35 shows an example of using the generic two-dimensional array to create a 1024x768 array of RGB values, all initialized to black.

Lists

This section shows how to implement dynamic one-dimension arrays, otherwise known as lists. The use of VHDL's dynamic memory allocation and *access* types is demonstrated. A technique similar to the one used for multi-dimensional arrays is used for Verilog. It concludes by the introduction of the *operator* concept.

Sample 4-35.
Example use
of the generic
two-dimen-
sional array

```
module testcase;
`include "rgb_typ.vh"
array_2 #(1024, 768, `RGB_SIZE) xvga();

initial
begin: test_procedure
    integer x,y;
    reg `RGB_TYP black;

    black`RED   = 8'd0;
    black`GREEN = 8'd0;
    black`BLUE  = 8'd0;

    for (x = 0; x < 1024; x = x + 1) begin
        for (y = 0; y < 768; y = y + 1) begin
            xvga.set(x, y, black);
        end;
    end
end
endmodule
```

Lists use memory
more efficiently
than arrays

Lists are similar to one-dimensional arrays. They are used to represent linear information. While the elements of an array can be
accessed randomly, the elements in a list must be accessed sequentially. If access time to various elements are your primary concern,
using an array is a more efficient implementation. On the other
hand, lists are more memory-efficient than arrays if not all locations
are used. Arrays must allocate memory for their entire size,
whereas lists can grow and shrink as the amount of information
they contain increases or decreases. If the memory usage of your
model is of concern, using lists may be the better approach.

Lists can be used
to model large
memories.

One of the best applications of a list is to model a large memory. In
system-level simulations, you may have to provide a model for a
large amount of memory. With the amount of memory available in
today's systems, and the overhead associated with modeling them,
you may find that you do not have a computer with enough
resources to simulate your system-level model efficiently. For
example, if you model a memory with 32-bits of addressable bytes
using an array of std_logic_vector in VHDL, the amount of
memory consumed by this array alone exceeds 128Gb (9 logic values requiring 4 bits to represent each std_logic bit x 8 bits per
byte x 4G).

Only the sections of the memory currently in use need to be modeled.

In any simulation, it is unlikely that all memory locations are required. Usually, the accesses are limited to a few regions within the memory address space. A list can be used to model a very large memory in a fashion similar to a cache memory. Only regions of the memory that are currently in use are stored in the list. When a particular location is accessed, the list is searched for the region of interest, allocating a new region as necessary.

Figure 4-4 shows a conceptual diagram of the various regions of memory within a list structure. This type of partial memory model is called *sparse memory*. The size of each individual region affects the ultimate performance of the simulation. With a smaller size, the memory is used more efficiently, but more regions are looked-up before finding the one of interest. Using larger regions has the opposite effect: more memory usage is traded-off for improved look-up efficiency.

Figure 4-4.
Sparse memory model

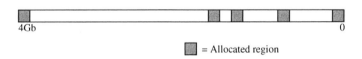

A linked list can be used to model a sparse memory.

A sparse memory model can be easily implemented using a list of records, where each record represents a region of the memory. The list can grow dynamically by allocating each region on demand, and linking each element in the list to another using access types. The list starts with a *head* access value that points to the first element in the list. Figure 4-5 shows a sparse memory model implemented using a *linked* list.

Figure 4-5.
Sparse memory model using a linked list

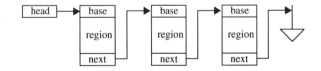

VHDL can implement a linked list.

The implementation of a sparse memory model using a linked list in VHDL[4] is shown in Sample 4-36. The memory regions are implemented as records: a field for the memory region itself (implemented as an array), and a field for the base address of that region. The record also contains a field to access the next region in the linked list. Because access types and access values are limited to variables, using such an implementation may be impractical if the list needs to be passed through interfaces.

Sample 4-36. Implementation of a sparse memory using a linked list in VHDL

```
process
   subtype byte is std_logic_vector(7 downto 0);
   type region_typ is array(0 to 31) of byte;

   type list_el_typ;
   type list_el_ptr is access list_el_typ;
   type list_el_typ is record
      base_addr  : natural;
      region     : region_typ;
      next_region: list_el_ptr;
   end record;

   variable head: list_el_ptr;

   -- See Sample 4-37 for continuation
begin
   ...
end process;
```

In Sample 4-37, a procedure is implemented to locate and return the section of the memory containing any address of interest. It starts at the head of the list, looking at every element of the list. If the end of the list is reached without finding the required region, a new region is allocated and prepended to the head of the list.

In Sample 4-38, a procedure is provided to read a single memory location. After locating the proper section of memory, it simply returns the content of the appropriate location in the section. There is also a procedure used to assign to a memory location. It works like the procedure to read a location, except that a new value is assigned.

Lists can be implemented using an array.

It may be necessary to use an array to implement a list. As mentioned earlier, access types and values in VHDL cannot be used for signals. If a list must be passed through the interface of an entity, a

4. For a more detailed description and alternative implementation, refer to section 6.1 of "*VHDL Answers to Frequently Asked Questions*", 2nd edition by Ben Cohen (Kluwer Academic Publisher, ISBN 0-7923-8115-7, 1998)

Sample 4-37.
Looking up a
sparse mem-
ory model in
VHDL

```
process
    -- See Sample 4-36 for declarations

    procedure get_region(addr: in  natural;
                         here: out list_el_ptr)
    is
        variable element: list_el_ptr;
    begin
        element := head;
        -- Have we reached the end of the list?
        while (element /= null) loop
            -- Is the address of interest in this
            -- list element?
            if (element.base_addr <= addr and
                addr < element.base_addr +
                       element.region'length) then
                here := element;
                return;
            end if;
            element := element.next_region;
        end loop;
        element := new list_el_typ;
        element.base_addr :=
            addr / element.region'length;
        element.next_region := head;
        head := element;
        here := element;
    end get_region;

    -- See Sample 4-38 for continuation
begin
    ...
end process;
```

different implementation strategy must be used. More importantly,
Verilog does not directly support dynamic memory allocation and
pointers or access values. There is a dynamic memory model PLI
package provided by Cadence in the distribution directory. You will
find it at:

$CDS_HOME/tools/verilog/examples/PLI/damem.

A sparse-memory
PLI package is
available.

This PLI package provides PLI routines that implement a sparse
memory model using hashed linked lists. However, if you wish to
use a list to model something other than a sparse memory (such as a
list of Ethernet packets to be applied to the design, or a list of
received video frames), you may want to implement them using

Sample 4-38.
Reading and
writing a loca-
tion in a sparse
memory
model

```
process
    -- See Sample 4-37 for declarations

    procedure lookup(addr : in  integer;
                        value: out byte) is
        variable element: list_el_ptr;
    begin
        get_region(addr, element);
        value := element.region(addr -
                                element.base_addr);
    end lookup;

    procedure set(addr:  in integer;
                    value: in byte) is
        variable element: list_el_ptr;
    begin
        get_region(addr, element);
        element.region(addr - element.base_addr) :=
            value;
    end set;

    variable val: byte;
begin
    set(10000, "01011100");
    lookup(10000, val);
    assert val = "01011100";
end process;
```

pure Verilog. Alternatively, using the *damem* package as an inspira-
tion, you could create a PLI package. But implementing a data
structure with PLI reduces the portability of the models using it.

Using an array to implement a list has an inherent restriction compared to a linked list implementation. Because an array has a fixed
length, a list implemented using such an array has a maximum
length equal to the array length. You have to provide an easy mean
of increasing the size of the array at compile-time to accommodate
simulations and testcases that require longer lists. In Sample 4-39,
you find an implementation of a generic list using an array. A parallel array is used to store a flag indicating if the corresponding array
element is a valid list element or not. The packets are records
implemented using the compiler symbol techniques described earlier in this chapter. The list is implemented as a module, just like the
multi-dimensional array. It is possible to have multiple instances of
this list.

<table>
<tr>
<td>

Sample 4-39.
Implementing
a list using an
array in Ver-
ilog

</td>
<td>

```verilog
module list;
parameter MAX_LEN = 100,
         IT_SIZE = 1;

reg [1:IT_SIZE] list_data [0:MAX_LEN];
reg [0:MAX_LEN] is_valid;

// Start with an empty list
initial is_valid = 0;

// See Sample 4-40 for continuation
endmodule
```

</td>
</tr>
</table>

Provide operators
for the data struc-
ture.

Lists are most useful when they come with a rich set of operators, such as appending or prepending to a list, removing the element at the head or tail of the list, finding out its length, or iterating over all of its elements. These operators should be provided in the same package as the data structure. The run-time efficiency of these operators is influenced by the storage policy of the list elements within the array. If the list is *packed*, i.e., all list elements are located in consecutive array locations at the beginning of the array, iterating over the content of the list is very efficient, but inserting or deleting elements in the middle of the list takes longer.

A *sparse* list, i.e., where list elements are separated by unoccupied array locations, are very efficient when deleting or inserting new elements, especially if the list is unordered. In Sample 4-40, there is a set of functions implementing an iterator in a sparse list. They can be used in *for-loop* statements to iterate over all of the elements of the list, as illustrated in Sample 4-41. The same module would also contain functions and tasks implementing the other operators, also called using hierarchical names.

Files

External input files
complicate config-
uration manage-
ment.

Personally, I prefer to avoid using external input files for test-benches. Configuration management of the testbench and the design under verification is complex enough. Without good practices, it is very difficult to make sure that you are simulating the right version of the correct model together with the proper implementation of the right testbench. If you must add to the mix making sure you have the right version of input files, often generated by scripts from some other format of some other files, configuration

Sample 4-40.
Implementing
a list iterator
in Verilog

```
module list;
// See Sample 4-39 for declarations

integer iterator;
reg     empty;

function [1:IT_SIZE] first;
   input dummy;
begin
   iterator = -1;
   first = next(0);
end
endfunction

function [1:IT_SIZE] next;
   input dummy;
begin: find_next
   iterator = iterator + 1;
   while (!is_valid[iterator] &&
          iterator <= MAX_LEN) begin
      iterator = iterator + 1;
   end
   empty = iterator > MAX_LEN;
   next  = list_data[iterator];
end
endfunction

endmodule
```

management grows exponentially in complexity. For example, many use files to initialize Verilog memories, as shown in Sample 4-42.

Understanding the implementation of the testcase now requires looking at two files and understanding their interaction. If the file always contains the same data for the same testcase, it can be replaced with an explicit initialization of the memory in the Verilog code, as shown in Sample 4-43. Now, only a single file needs to be managed and understood.

VHDL has primi-
tive input/output
while Verilog as
strong output, but
poor input.

VHDL has a general-purpose, albeit very primitive, file input and output capability. At the time this book was written, Verilog had a very strong file output capability, but its file input features were almost non-existent. A standard system task similar to C's *fscanf* routine is being proposed for inclusion in the next version of the standard. For those of you who cannot wait, more information

<table>
<tr>
<td>

Sample 4-41.
Using a list
iterator in Ver-
ilog

</td>
<td>

```verilog
module testcase;
`include "packet.vh"

list #(32, `PACKET_SIZE) packet_list();

initial
begin: test_procedure
    reg `PACKET_TYP packet;
    integer last_seq;
    // Receive packets
    while (...) begin
        ...
        packet_list.append(packet);
    end
    // Verify that packets were received in order
    last_seq = -1;
    for (packet = packet_list.first(0);
         !packet_list.empty;
         packet = packet_list.next(0)) begin
        if (packet`SEQ_ID >= last_seq) ...
        last_seq = packet`SEQ_ID;
    end
end
endmodule
```

</td>
</tr>
<tr>
<td>

Sample 4-42.
Initializing a
Verilog mem-
ory using an
external file.

</td>
<td>

```verilog
module testcase;

reg [7:0] pattern [0:55];

initial $readmemh(pattern, "pattern.memh");

endmodule
```

</td>
</tr>
<tr>
<td>

Sample 4-43.
Explicitly ini-
tializing a Ver-
ilog memory

</td>
<td>

```verilog
module testcase;

reg [7:0] pattern [0:55];

initial
begin
    pattern[0]  = 8'h00;
    pattern[1]  = 8'hFF;
    ...
    pattern[55] = 8'hC0;
end

endmodule
```

</td>
</tr>
</table>

about a public-domain implementation of a similar task using PLI can be found in the *resources* section at:

http://janick.bergeron.com/wtb

Verilog can only read binary and hexadecimal values.

The current version of Verilog can only read files of binary or hexadecimal values into a memory. If you want to provide high-level data to Verilog via files, you have to "compile" it into low-level numerical values, then interpret this low-level form back into high-level data inside Verilog.

If the external file is automatically generated from some other file, a different approach can be used to circumvent Verilog's limitation. Instead of generating data, why not generate Verilog code? The intermediate form is much easier to debug than an artificial numerical form coupled with translation and interpretation steps. The code shown in Sample 4-43 could have been just as easily generated as the data file used by Sample 4-42. The technique can be used with VHDL as well, helping to circumvent the primitiveness of its file input features.

External files can eliminate recompilation.

Using external input files can save a lot of simulation time if you use a compiled simulator such as *NC-Verilog*, *VCS*, or any VHDL simulator. If you can modify your testcase by modifying external input files. It is not necessary to recompile the model of the design under verification nor the testbench. For large designs, this compilation time can be significant, especially for a gate-level design with SDF back-annotation.

Files can program bus-functional models.

Programmable testbenches are architected around programmable bus-functional models and checkers, and programmed via an external input file. The "program" can be as simple as a sequence of data patterns or as complex as a pseudo assembly language with opcodes and operands interpreted by an engine implemented in Verilog or VHDL.

Interfacing High-Level Data Types

It is very unlikely that high-level data types are directly usable by any device that must be verified. Any complex data structure has to be sent to or received from the design using a simpler protocol implemented at the bit level, usually including synchronization, framing, or handshaking signals. In Chapter 5, I show techniques using bus-functional models for applying data to a design from high-level data structures (and vice-versa on the output side).

THE HDL PARALLEL ENGINE

C and C++ lack essential concepts for hardware modeling.

Why hasn't C been used as a hardware description language instead of creating Verilog and VHDL (and many other proprietary ones)? Because C lacks three fundamental concepts necessary to model hardware designs: connectivity, time, and concurrency.

Connectivity, Time, and Concurrency..

Connectivity is the ability of describing a design using simpler blocks then connecting them together. Schematic capture tools are perfect example of connectivity support

Time is the ability to represent how the internal state of a design evolves over time. This concept is different from *execution time* which is a simple measure of how long a program runs.

Concurrency is the ability to describe actions that occur at the same time, independently of each other.

Given enough time and creativity, you could *model* these concepts in a C program. However, these basic concepts would have to be implemented over and over in a customized fashion for each design. It is much more efficient to use a language where these concepts are built-in

C and C++ could be extended.

There have been many attempts to extend C or C++ to include some or all of these concepts. We may see some extended versions of C++ promoted as a verification language since it provides excellent support for data abstraction, encapsulation and object-oriented design. For example, the *VERA* verification language feels like a hybrid between Verilog and C++ and the *systemC* initiative is an open-source initiative to develop a C++-based design and verification environment. More information on the *systemC* initiative can be found in the *resources* section of:

http://janick.bergeron.com/wtb

Connectivity, Time, and Concurrency in HDLs

Verilog and VHDL implement these concepts in different ways.

The connectivity, time, and concurrency concepts are very important to understand when learning to model using a hardware description language. Each language implements them in a different fashion, some easier to understand than the other.

For example, connectivity in Verilog is implemented by directly instantiating modules within modules, and connecting the pins of the modules to wires or registers. Understanding why registers cannot be used in some circumstances requires understanding the concept of concurrency. Concurrency is described in detail in the following sections.

In VHDL, connectivity is implemented with entities, architectures, components, and configurations. The mechanics of connectivity in VHDL require a lot of statements and apparent duplication of information and is often one of the most frustrating concept to learn in VHDL.

Verilog implements time as unitless relative values.

The concept of time is also implemented differently. Verilog uses a unit-less time value. The time values from multiple modules are correlated using a scale factor specified using the `timescale compiler directive. In VHDL, all time values are absolute, with their units clearly stated.

VHDL and Verilog differ most in their implementation of concurrency.

The implementation of concurrency is where VHDL and Verilog differ the most. Although they both use an *event-driven* simulation process, they differ in the granularity of their concurrency, and in the timing and focus of assignments between concurrent constructs. To write VHDL, it is necessary to understand the implementation of this concept because of the restrictions concurrency imposes on the use of the language. Verilog's implementation puts very few restrictions on the use of the language. Verilog relies on the designer to use concurrency appropriately. If you are limited to a certain coding style, such as the synthesizeable subset, you can write functional Verilog code without having to understand its implementation of the concept of concurrency.

You write better testbenches when you understand concurrency.

When writing testbenches, you are not confined to such coding styles. It becomes necessary to understand how concurrency is implemented and how concurrency affects the execution of the various components of the testbench.

Many testbenches are written with a severe lack of understanding of this important concept. In the best case, the execution and overall control structure of the testbench code is difficult to follow and maintain. In the worst case, the testbench fails to execute properly on a different simulator, on different versions of the same simulator, or when using different command-line options. The understanding of concurrency is often what separates the experienced designer from the newcomers.

The Problems with Concurrency

There are two problems with concurrency. The first one is in describing concurrent systems. The second is executing them.

Concurrent systems are difficult to describe.

Since computers were created, computer scientists have tried to figure out a way to take advantage of the increased performance offered by multi-processor machines. They are relatively easy to build and many parallel architectures have been designed. However, they proved much more difficult to program. I do not know if that difficulty originated with the mindset imposed by the early Von Neumann architecture still used in today's processors, or by an innate limitation of our intellect.

Concurrent systems are described using a hybrid approach.

Human beings are adept at performing relatively complex tasks in parallel. For example, you can drive in heavy traffic while carrying a conversation with a passenger. But it seems that we are better at describing a process or following instructions in a sequential manner. For example, a recipe is always described using a sequence of steps. The description of concurrent systems has evolved into a hybrid approach. Individual processes running in parallel with each other are themselves described using sequential instructions. For example, a desert recipe includes instructions for the cake and the icing as separate instructions that can be performed in parallel, but the instructions themselves follow a sequential order.

VHDL and Verilog models are concurrent processes described sequentially.

A similar principle is used in both VHDL and Verilog. In VHDL, the concurrent processes are the *process* statements (all concurrent statements are simple short-hand forms for processes). In Verilog, the concurrent processes are the *always*, and *initial* blocks and the *continuous signal assignment* statements. The exact behavior of each instance of these constructs, in both languages, is described individually using sequential statements.

Every *process* in a VHDL model, and every *always* and *initial* block in a Verilog model execute in parallel with each other, but internally each executes sequentially. It is a common misconception that Verilog's *initial* blocks mean "initialize". Unlike VHDL, there is <u>no</u> initialization phase on Verilog. Everything is implicitly initialized to 'x' and *initial* blocks are identical to *always* blocks except that they execute only once. They are removed from the simulation once the last statement in the *initial* block executes.

Concurrent systems must be executed on single processor machines. The second problem with concurrency is in executing it. If you look inside the workstation that you use to simulate your VHDL or Verilog model, you will see that there is a single processor. Even if you had access to a multi-processor machine, you can always write a model with one more parallel constructs than you have processors available. How do you execute a parallel description on a single processor, which is itself a sequential machine?

Emulating Parallelism on a Sequential Processor

Multi-tasking operating systems are like simulators. If you use a modern computer, you probably have a windows-based graphical interface. During normal day-to-day use, you are very likely to have several windows open at once, each of them running a different application. On multi-user machines, there may be several others running a similar environment on the same computer. The applications running in all of these windows appear to work all in parallel even though there is a single sequential processor to execute them. How is that possible? You probably answered *time-sharing*. With time-sharing, each application uses the entire processor for small portions of time. Each application has its turn according to priority and activity. If the performance of the processor and operating system is high enough, the interruptions in the execution of a program are below our threshold of detection: it appears as if each program runs smoothly 100 percent of the time, in parallel with all the others.

Simulators are time-sharing engines. A VHDL or Verilog simulator works using the same principle. Each *process*, or *always* and *initial* block has the simulation engine for some portion of time, in turn, one at a time. Each *appears* to be executing in parallel with the others when, in fact, they are each executed sequentially, one after another. There is one important difference in the time-sharing process of a simulator. Unlike a multi-tasking operating system, it assumes that the various parallel constructs cooperate to obtain fair access to the simulation resources.

Simulators do not have time slice limits.

In an operating system, every process has a limit on the amount of processor time it can have during each execution slice. Once that limit is exhausted, the process is kicked out of the processor to be replaced by another. There is no such limit in a simulator. Any process keeps executing until it explicitly requests to be kicked out. It is thus possible, in a simulation, to have a process grab the simulation engine and never let it go. Ensuring that the parallel constructs properly cooperate in a simulation is a large part of understanding how concurrency is implemented.

Processes simulate until they execute a *wait* statement.

In VHDL, a process simulates, and keeps simulating, until a *wait* statement is executed. When the *wait* statement is executed, the process is kicked out of the simulation engine and replaced by another one. This process remains "out of circulation" until the condition it is waiting for is realized. Verilog has a similar model: *always* and *initial* blocks simulate and keep simulating until a @, #, or a blocking assignment is executed. It also stops executing if a *wait* statement whose condition is currently <u>false</u> is executed. If a *process* or an *always* or *initial* block does not execute some form of a *wait* statement, it remains in the simulation engine, locking all other processes out.

The Simulation Cycle

Simulators execute processes at the current time, then assign zero-delay future values.

Figure 4-6 shows the VHDL and Verilog simulation cycle. For a given timestep, the simulation engine executes each of the parallel processes that must be executed. While executing, these processes may perform assignments of future values using signal assignments in VHDL or non-blocking assignments in Verilog. Once all pro-

Figure 4-6.
VHDL and
Verilog
simulation
cycle

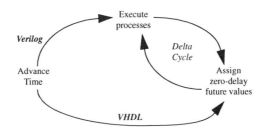

cesses are executed (i.e., they are all waiting for something), the simulator assigns any future values scheduled for the current timestep (i.e. zero-delay assignments). Processes sensitive to the new values are then executed at the next timestep. This cycle continues until there are no more processes that must be executed at the current timestep and there are no more zero-delay future values.

Simulators then advance time or starve.

If there is nothing left to be done at the current time, there <u>must</u> be either:

1. A process waiting for a specific amount of time

2. A future value to be assigned after a non-zero delay.

If neither of the conditions are true, then the simulation stops on its own, having reached a quiescent state and suffering from event starvation. If one of the conditions is present, the simulator advances time to the next time period where there is useful work to be done. The simulator then assigns a future value, which causes processes sensitive to the signals assigned these values to be executed, or execute processes that were waiting.

Simulators do not increment time step by step.

Notice that the simulator does <u>not</u> increment time by a basic time unit, timestep, or time increment. Regardless of the simulation resolution, the simulation advances time as far as necessary, in a single step, to the next point in time where there is useful work to do. Usually, that point in time is the delay in the clock generator. Increasing the simulation time resolution should not significantly decrease the simulation performance of a behavioral or RTL model.

Zero-delay cycles are called *delta cycles*.

The state of the simulation progresses along two axis: zero-time and simulation time. As processes are simulated and new values are assigned after zero delays, the state of the simulation evolves and progresses, but time does not advance. Since time does not advance, but the state of the simulation evolves, these zero-delay cycles where processes are evaluated and zero-delay future values are assigned, are called *delta-cycles*. The simulation progresses first along the delta axis then along the real-time axis, as shown in Figure 4-7. It is possible to write models that simulate entirely along the delta axis. It is also possible to write models that are unintentionally stuck in delta cycles, preventing time from advancing.

Figure 4-7.
Time
progression
along two axis

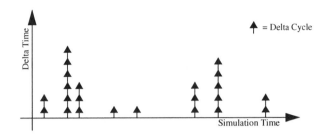

VHDL and Verilog
behave differently
after advancing
time.

In Figure 4-6, you will notice that the VHDL and Verilog simulation cycles differ after time advances. In VHDL, future values are
assigned before the execution of processes. In Verilog processes are
executed first. Given the choice between executing a process or
assigning a new value at the exact same point in time in the future,

Sample 4-44.
Verilog model
apparently
identical to a
VHDL model

```
module testcase;

reg R;

initial
begin
    R = 1'b0;
    R <= #10 1'b1;
    #10;
    if (R !== 1'b1) $write("R is not 1\n");
end

endmodule
```

Sample 4-45.
VHDL appar-
ently identi-
cal to a Ver-
ilog model

```
entity case is
end case;

architecture test of case is
    signal R: bit := '0';
begin
    process
    begin
        R <= '1' after 10 ns;
        wait for 10 ns;
        assert R = '1'
            report "R is not 1"
            severity NOTE;
        wait;
    end process;
end test;
```

VHDL assigns the new value, while Verilog executes the process first. This may produce different simulation results between apparently identical VHDL and Verilog models, such as those shown in Sample 4-44 and Sample 4-45. A message is displayed in the Verilog version, but not in the VHDL one. It may also affect the behavior in a co-simulation environment when a new value crosses the VHDL/Verilog boundary.

Parallel vs. Sequential

Use sequential descriptions as much as possible.

As explained earlier, humans can understand sequential descriptions much easier than concurrent descriptions. Anything that is described using a single sequence of statements is easier to understand and maintain than the equivalent behavior described using parallel constructs. The independence of their location and ordering in the source file adds to the complexity of concurrent descriptions. A concurrent description that would be relatively easy to understand can be obfuscated by simply separating the pertinent concurrent statements with a few other unrelated concurrent constructs. Therefore, functionality should be described using sequential constructs as much as possible.

A frequent misuse of sequential constructs in Verilog involves the initialization of registers. For example, Sample 4-46 shows a clock generator implemented using two concurrent constructs: an *initial* and an *always* block.

Sample 4-46.
Misuse of concurrency in Verilog

```
reg clk;
initial clk = 1'b0;
always #50 clk = ~clk;
```

However, generating a clock is an inherently sequential process: it starts at one value then toggles between one and zero at a constant rate. A better description, using a single concurrent construct, is shown in Sample 4-47.

Deterministic sequential behavior does not need concurrency.

Another, less obvious, case of misused concurrency happens when the behavior of the various processes is deterministically sequential because of the data flow. For example, Sample 4-48 shows a VHDL process labeled *P2* that can execute only once the process labelled *P1* triggers the signal *do*. The *P1* process then waits for the comple-

Sample 4-47.
Proper use of
concurrency in
Verilog

```verilog
reg clk;
initial
begin
   clk = 1'b0;
   forever #50 clk = ~clk;
end
```

tion of process *P2* before resuming its execution. The sequence of execution cannot be other than the first half of *P1*, *P2*, then the second half of *P1*.

Sample 4-48.
Deterministic
sequential exe-
cution in
VHDL

```vhdl
architecture test of bench is
   signal do, done: boolean;
begin
   P1: process
   begin
      -- First half of P1
      . . .
      do <= not do;
      wait on done;
      -- Second half of P1
      . . .
   end process P1;

   P2: process
   begin
      wait on do;
      -- All of P2
      . . .
      done <= not done;
   end process P2;
end test;
```

Sample 4-49.
Simplified
sequential exe-
cution in
VHDL

```vhdl
architecture test of bench is
begin
   P1_2: process
   begin
      -- First half of P1
      . . .
      -- All of P2
      . . .
      -- Second half of P1
      . . .
   end process P1_2;
end test;
```

The implementation in Sample 4-49 shows the equivalent functionality, implemented using a single process. Not only is the execution flow easier to follow, but it does not require the control signals *do* and *done*.

Fork/Join Statement

Control flow may alternate between sequential and concurrent regions.

The overall control flow for a testcase often involves a sequence of sequential steps followed by concurrent ones. For example, testing a configuration of a design may require configuring the device through several consecutive reads and writes via the CPU interface, then concurrently sending and receiving data. This process is then repeated for another configuration. Figure 4-8 shows a control flow diagram of such a control structure.

Figure 4-8.
Series of sequential and concurrent control flows

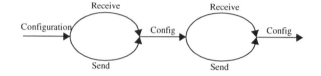

Implement using a *fork/join* statement in Verilog.

The easiest way to implement this type of control flow structure is to use a *fork/join* statement in Verilog. This statement dynamically creates concurrent processes within a region of sequential code. The sequential execution resumes after the *join* statement, once all the concurrent regions are complete. For example, the code in Sample 4-50 waits for the maximum of Ta, Tb, and Tc.

Sample 4-50.
Example of using the *fork/ join* statement in Verilog

```
initial
begin
    ...
    fork
        #(Ta);
        #(Tb);
        #(Tc);
    join
    ...
end
endmodule
```

The *fork/join* statement can be disabled.

It is often necessary to have the *fork/join* statement continue the sequential execution after only one of the concurrent regions has completed. This can be accomplished by disabling the <u>named *fork/*</u>

join statement from within the execution branches. For example, the code in Sample 4-51 detects and reports a time-out if the *posedge* on *gt* is not received within *Tmax* time units.

Sample 4-51.
Example of
disabling the
fork/join state-
ment in Ver-
ilog

```
initial
begin
    . . .
    fork: wait_for_gt
        @ (posedge gt) disable wait_for_gt;
        #(Tmax) begin
            $write("Time-out on gt\n");
            disable wait_for_gt;
        end
    join
    . . .
end
endmodule
```

VHDL has no
fork/join construct.

Unfortunately, VHDL does not have a *fork/join* statement or its equivalent. It is necessary to emulate this behavior using separate processes and controlling their execution via another process. Emulating the functionality of the *fork* is simple: an event on a single signal can be used to trigger the execution of the concurrent regions. Emulating the functionality of the *join* is more compli-cated. You could use an event on a signal for each branch of the join, but this would require a signal for every branch. Adding a new branch would require adding a new signal and modifying the *wait* statement implementing the *join*.

Sample 4-52.
Emulation of
the *fork/join*
statement in
VHDL

```
package fork_join is

type join_ctl_typ is (join, fork, run);
type branches_typ is
    array(integer range <>) of join_ctl_typ;

function join_all(branches: branches_typ)
    return join_ctl_typ;
function join_one(branches: branches_typ)
    return join_ctl_typ;

subtype fork_join_all is join_all join_ctl_typ;
subtype fork_join_one is join_one join_ctl_typ;

end fork_join;
```

Emulate the *join* statement using a resolution function.

Resolution functions provide a simpler mechanism for handling an arbitrary number of branches. A different resolution function can be used to implement a *join-all* or a *join-one* functionality. In a *join-all*, all branches of the *fork* must complete for the *join* to complete. In a *join-one*, any single branch, once completed, causes the *join* to complete. Sample 4-52 and Sample 4-53 show the implementation of the *join* resolutions functions, while the code in Sample 4-54 shows how to use it. My prayers to the VHDL gods for a *fork/join* statement remain, to this day, unanswered.

Sample 4-53. Implementation of the *fork/join* emulation in VHDL

```
package body fork_join is

function join_all(branches: branches_typ)
    return join_ctl_typ is
begin
    for I in branches'range loop
        if branches(i) = fork then
            return fork;
        end if;
        if branches(i) = run then
            return run;
        end if;
    end loop;
    return join;
end join_all;

function join_one(branches: branches_typ)
    return join_ctl_typ is
begin
    for I in branches'range loop
        if branches(i) = fork then
            return fork;
        end if;
        if branches(i) = join then
            return join;
        end if;
    end loop;
    return run;
end join_one;

end fork_join;
```

Sample 4-54.
Using the
emulation of
the *fork/join*
statement in
VHDL

```
use work.fork_join.all;
architecture test of bench is
    signal fk_jn1: fork_join_all;
begin
    process
    begin
        -- Fork
        fk_jn1 <= fork;
        wait until fk_jn1 = fork;
        fk_jn1 <= run;
        -- Branch #0

        . . .
        fk_jn1 <= join;
        -- Join
        wait until fk_jn1 = join;
    end process;

    branch1: process
    begin
        fk_jn1 <= fork;
        wait until fk_jn1 = fork;
        fk_jn1 <= run;

        . . .
        fk_jn1 <= join;
        wait;
    end process branch1;

    branch2: process
    begin
        fk_jn1 <= fork;
        wait until fk_jn1 = fork;
        fk_jn1 <= run;

        . . .
        fk_jn1 <= join;
        wait;
    end process branch2;
end test;
```

The Difference Between Driving and Assigning

Assignments write
a value to a mem-
ory location.

Regular programming languages provide variables that can contain
arbitrary values of the appropriate type. They are implemented as
simple memory locations. Assigning to these variables is the simple
process of storing a value into that memory location. VHDL variables and Verilog registers operate in the same way. When an
assignment is completed, whether blocking or non-blocking, the
newly assigned value overwrites any previous value in the memory

location. Previous assignments have no effects on the final result. Regular assignments behave like a multiplexer. A single value from all of the potential contributors is somehow selected.

The last assign-
ment determines
the value.

For example, in Sample 4-55, the value of the register *R* goes from 'x' to 5 to 4 to 3 to 2 to 1, then finally to 0. Since *R* is a variable shared by all three concurrent blocks, a single memory location exists. Whatever value was assigned last, by a concurrent block, is the value stored in the variable. This is where the name *register* comes from for Verilog variables. Registers - or flip-flops - retain whatever value was last loaded into them, without regard to the previous values or other concurrent sources.

Sample 4-55.
Assignments
to a shared
variable in
Verilog

```
module assignments;
integer R;

initial R <= #20 3;

initial
begin
   R = 5;
   R = #35 2;
end

initial
begin
   R <= #100 1
   #15 R = 4;
   #220;
   R = 0;
end

endmodule
```

Hardware descrip-
tion languages
need the concept
of a wire.

The variable is sufficient for ordinary sequential programming languages. When describing hardware, a construct that can describe the behavior of a wire used to connect multiple devices together must be provided. Figure 4-9 shows a wire, presumably part of a data bus, connected to several devices. Each device, using a tristate driver, can drive a value onto the wire. The final logic value on the wire depends on *all* the individual values being driven, not just the last one, like a variable.

Figure 4-9.
Multiple
drivers on a
wire

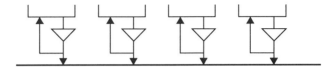

Individual values
from connected
devices must be
continuously
driven onto the
wire.

To properly model connectivity via a wire, any value driven by a device must be continuously driven onto that wire, in parallel with the other driving values. The final value on that wire depends on all of the continuously driven individual values.

For example, on a tristate wire, the individual driven values of 'z', '1', 'weak-0' and 'z' would produce a final result of '1'. Figure 4-10 shows the implementation of the wire driver in each language.

In Verilog, this continuous drive is implemented using a continuous assignment while the final value is determined by the type of wire being used: *wire*, *wor*, *wand*, or *trireg*.

In VHDL, the continuous drive is implemented in each process that assigns a signal while the final value is determined by the user-defined resolution function.

Figure 4-10.
Implementa-
tion of
continuous
drive in
Verilog and
VHDL

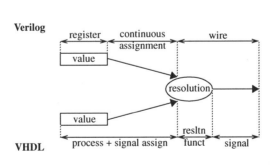

Each concurrent
construct has its
own, single driver.

Parallel drivers on a wire require concurrent constructs to describe them. Many inexperienced engineers, when learning to code for synthesis try to implement the design shown in Figure 4-11 using the code shown in Sample 4-56. Unfortunately, since a single register is used with variable assignments in sequential code, a *multi-*

plexor is synthesized instead of the expected parallel drivers. The proper solution requires three concurrent constructs, one for each driver, and is shown in Sample 4-57.

Figure 4-11.
Simple design with three tristate drivers

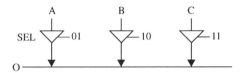

Sample 4-56.
Implementation using a multiplexer

```
module simple(A, B, C, SEL, O);
input       A, B, C;
input  [1:0] SEL;
output      O;

reg O;
always @ (A or B or C or SEL)
begin
   case (SEL)
   2'b00: O = 1'bz;
   2'b01: O = A;
   2'b10: O = B;
   2'b11: O = C;
   endcase
end
endmodule
```

Sample 4-57.
Implementation using three tristate drivers

```
module simple(A, B, C, SEL, O);
input       A, B, C;
input  [1:0] SEL;
output      O;

assign O = (SEL == 2'b01) ? A : 1'bz;
assign O = (SEL == 2'b10) ? B : 1'bz;
assign O = (SEL == 2'b11) ? C : 1'bz;

endmodule
```

VERILOG PORTABILITY ISSUES

Two compliant simulators can produce different results.

In my many years of consulting in design verification, I have yet to see a <u>single</u> testbench that simulates with identical results on *Verilog-XL* and *VCS*. Half the time, these same testbenches can produce different results by using different command-line options! Yet, both Verilog simulators are fully compliant with the IEEE standard. Sometimes, the problem lies with the standard: many implementation details were left unspecified or existing discrepancies between simulators were also declared "unspecified". The other and bigger problem lies with the authors and their lack of understanding of concurrency and how race conditions are created.

The problem of unspecified behavior or race conditions is conveniently eliminated when limiting yourself to writing synthesizeable code. But once you start using all the features of the language, you may find yourself with code that is not portable across different simulators. Verilog *appears* easy to learn because it produces the expected response rather quickly. Making sure that the results are reproducible under different conditions is another matter. Learning the idiosyncrasies of the language are what takes time and differentiates an experienced modeler from a new one.

Shared variables in VHDL can create race conditions.

VHDL was designed to make race conditions impossible to implement. However, with the introduction of shared variables in the 1993 version of the standard, race conditions are now just as easily introduced in VHDL as in Verilog. If you use shared variables in VHDL, pay close attention to the race conditions described below.

Read/Write Race Conditions

A *read/write* race condition happens when two concurrent blocks attempt to read and write the same register in the same timestep. If you look at the code in Sample 4-58, you will notice that the first *always* block assigns the register *count* while the second one displays it. But both blocks execute at the rising edge of the clock.

The execution order determines the final result.

If you refer to Figure 4-6 and the section titled "Emulating Parallelism on a Sequential Processor" on page 128, you will see that both blocks are executed one after another, during the same timestep. The order in which the blocks are executed is not deterministic. Let's assume that the current value of *count* is 10. If the first block is executed first, the value of *count* is updated to 11. When the second block is executed, the value 11 is displayed. However, if the second block executes first, the value of 10 is displayed, the value of *count* being incremented only when the first block executes later.

<table>
<tr>
<td>

Sample 4-58.
Example of a
read/write race
condition

</td>
<td>

```
module rw_race(clk);
input clk;

integer count;

always @ (posedge clk)
begin
    count = count + 1;
end

always @ (posedge clk)
begin
    $write("Count is equal to %0d\n", count);
end

endmodule
```

</td>
</tr>
</table>

Some read/write
race conditions
can be solved by
using non-block-
ing assignments.

This type of race condition can be easily solved by using a non-blocking assignment, such as shown in Sample 4-59. Referring again to Figure 4-6: when the first block executes, the non-blocking assignment *schedules* the new value of 11, with a delay of zero, to the next timestep. When the second block executes, the value of *count* is <u>still</u> 10. The new value is assigned to *count* only when all blocks executing at this timestep are executed, creating a delta cycle.

<table>
<tr>
<td>

Sample 4-59.
Avoiding a
read/write race
condition

</td>
<td>

```
module rw_race(clk);
input clk;

integer count;

always @ (posedge clk)
begin
    count <= count + 1;
end

always @ (posedge clk)
begin
    $write("Count is equal to %0d\n", count);
end

endmodule
```

</td>
</tr>
</table>

A more insidious read/write race condition can occur between *always* or *initial* blocks and continuous assignments. Examine the code in Sample 4-60 closely. What value of *out* will be displayed?

Sample 4-60.
A Verilog riddle

```
module rw_race;

wire [7:0] out;
assign out = count + 1;

integer count;
initial
begin
    count = 0;
    $write("Out = %b\n", out);
end

endmodule
```

The answer is "xxxxxxxx" or "00000001".

The answer depends on the simulator and the command line you are using. Without using any command-line options, *Verilog-XL* says that *out* is "xxxxxxxx". *VCS* says that *out* is "00000001". Why the difference of opinion?

Verilog-XL does not interrupt blocks to execute continuous assignments.

The difference comes from their interpretation of the simulation cycle. When the *initial* block assigns a new value to *count*, *Verilog-XL* schedules the execution of the continuous assignment for the next timestep, since it is sensitive to *count*. The execution of the *initial* block is not interrupted and the value of *out* displayed is the one it had after initialization, since the continuous assignment has not yet been executed.

VCS does.

VCS, on the other hand, executes the continuous assignment <u>as soon as *count* is assigned</u> in the *initial* block. The execution of the *initial* block is interrupted after the assignment to *count* while the continuous assignment is executed. The execution of the *initial* block resumes immediately afterward. The immediate propagation of events through continuous assignments is one of the techniques *VCS* implementers have used to speed-up simulation, unfortunately at the price of incompatibility with *Verilog-XL*.

This type of race condition cannot be easily avoided.

Unfortunately, this type of error condition is not as easy to avoid or eliminate as the one between two blocks. When writing behavioral code, you must be careful about the timing between assignments to registers in the right-hand side of a continuous assignment, and reading the wire driven by it. To make matters worse, the race condition may involve non-zero delays as well as multiple continuous assignment statements, such as in Sample 4-61. A *read/write* race condition occurs if the delay between the time the right-hand side of a continuous assignment is updated, and the time any wire on the left-hand side is read, is equal to the propagation delay of all intervening continuous assignments. Figure 4-12 illustrates the timing of these race conditions.

Sample 4-61.
Another read/
write race condition

```
module rw_race;

wire [7:0] out, tmp;
assign #1 out = tmp - 1;
assign #3 tmp = count + 1;

integer count;
initial
begin
   count = 0;
   #4;
   // "out" will be 0 or x's.
   $write("Out = %b\n", out);
end

endmodule
```

Figure 4-12.
Timing of a
read/write
race condition

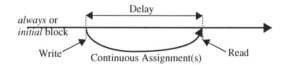

Write/Write Race Conditions

A *write/write* race condition occurs when two *always* or *initial* blocks write to the same register at the same timestep. If you look at the code in Sample 4-62, you will notice that both *always* blocks assign the register *count* and both blocks execute at the rising edge of the clock.

The execution order determines the final result.

If you refer one more time to Figure 4-6 and the section titled "Emulating Parallelism on a Sequential Processor" on page 128, you will see that both blocks are executed one after another, during the same timestep. Again, the order in which the blocks execute is

Sample 4-62.
Example of a
write/write
race condition

```
module ww_race(clk);
input clk;

integer count;

always @ (posedge clk)
begin
   count = count + 1;
end

always @ (posedge clk)
begin
   count = count / 2;
end

endmodule
```

<u>not deterministic</u>. Let's assume that the current value of *count* is 10. If the first block is executed first, the value of *count* is updated to 11. When the second block is executed, the value of *count* is updated to 5. However, if the second block executes first, the value of *count* is updated to 5, then incremented to 6 when the first block executes later.

Sample 4-63.
Another exam-
ple of a write/
write race con-
dition

```
module ww_race(clk);
input clk;

integer count;

always @ (posedge clk)
begin
   count <= count + 1;
end

always @ (posedge clk)
begin
   count <= count / 2;
end

endmodule
```

Non-blocking
assignments do not
solve the problem.

You might be tempted to use the same solution to eliminate the race condition as was used to eliminate the *read/write* race condition, as shown in Sample 4-63. Using non-blocking assignments simply moves the *write/write* race condition from the register assignment to the scheduling of the future value. If the first block executes first, the future value 11 is scheduled for the next timestep. When the second block executes, the future value 5 is also scheduled for the next timestep, overwriting the previously scheduled value of 11. If the blocks execute in the opposite sequence, the scheduled value of 11 overwrites the previously scheduled value of 5.

Pop quiz!

Why can't you have a *write/write* race condition on a wire?[5]

Initialization Races

There is no initial-
ization phase in
Verilog.

The most frequent race conditions can be found at the beginning of the simulation, when all blocks are executed for the first time. Unlike VHDL, Verilog has <u>no</u> initialization phase. Everything is initialized to 'x', then the simulation starts normally. It is a common misconception that *initial* blocks are used to initialize variables. *Initial* blocks are <u>identical</u> to *always* blocks, except that they execute only once, whereas *always* blocks execute forever, as if they were stuck in an infinite loop.

Initial blocks are
not executed first.

When the simulation is started, the *initial* and *always* blocks are executed one after another, in <u>any order</u>. The *initial* blocks are <u>not</u> executed first - although doing so would not be illegal and some simulators, such as *Silos III*, do just that. Most simulators, for no other reason than to be compatible with *Verilog-XL* and legacy code containing race conditions, first execute blocks in the same order as they are specified in the file. But subsequent execution order is not so deterministic.

When simulating the code in Sample 4-64 using an XL-compliant simulator, the first *always* block would be executed and interrupted immediately, waiting for the rising edge of the clock. The *initial* block is executed next, assigning the new value of '1' to the register named *clk*, which was previously initialized to 'x'. A transition from 'x' to '1' being considered a rising edge, the first *always* block sees the event and is scheduled to be executed again at the next timestep. However, since the last *always* block was <u>not</u> yet executed, and thus is not waiting for the rising edge of the clock, it does not see this edge. When the last block is finally executed, it is

---

5. Because wires are driven, not assigned. The value from each parallel construct would contribute to the final logic value on the wire, without overwriting the other.

also immediately suspended, waiting for the <u>next</u> rising edge on *clk*. An XL-compliant simulator would therefore execute the body of the first *always* block, but not the second. However, that is not a requirement. If a simulator chooses to execute the *initial* block first, the body of neither block would execute at time 0.

Sample 4-64.
Race condition
at simulation
startup

```
module init_race;
reg clk;

always @ (posedge clk)
begin
    $write("Block #1 at %t\n", $time);
end

initial clk = 1'b1;

always @ (posedge clk)
begin
    $write("Block #3 at %t\n", $time);
end

endmodule
```

Guidelines for Avoiding Race Conditions

Race conditions can be avoided if you follow strict coding guidelines, effectively restricting the usage of Verilog to what is automatically enforced by VHDL.

1. If a register is declared outside of the *always* or *initial* block, assign to it using a non-blocking assignment. Reserve the blocking assignment for registers local to the block.

2. Assign to a register from a single *always* or *initial* block.

3. Use continuous assignments to drive inout pins only. Do not use them to model internal combinatorial functions. Prefer sequential code instead.

4. Do not assign any value at time 0.

Events from Overwritten Scheduled Values

If a scheduled value is overwritten by another scheduled value, can the original value cause an event? The answer to that question is left undefined by the Verilog standard. If you look at the code in Sample 4-65, will anything be displayed at time 10?

Sample 4-65.
Overwriting scheduled values in Verilog

```
module events;

reg stobe;

always @ (strobe)
begin
    $write("Stobe is %b\n", strobe);
end

initial
begin
    strobe = 1'b0;
    strobe <= #10 1'b1;
    strobe <= #10 1'b0;
end

endmodule
```

Overwriting a scheduled value may generate an event.

Figure 4-13 shows the queue of scheduled future values for register *strobe* just before the last statement of the *initial* block is about to execute. After executing that last statement, and scheduling the new value of '0' after 10 time units in the future, what happens to the previously scheduled value of '1'? Is it removed? Is it left there? If so, which value will be assigned to *strobe* 10 time units from now? Only '0' (and thus not generating an event on *strobe*) or both in zero-time (and generating an event)? The answer to this question is simulator dependent. Avoid overwriting previously scheduled values using non-blocking assignments.

Figure 4-13.
Event queue on *strobe*

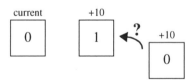

Disabled Scheduled Values

Disable statements can be used to control loops.

The *disable* statement is great for modeling reset conditions (see "Modeling Reset" on page 276 for more details) or loop control to emulate the behavior of VHDL's *next* and *exit* statements. The code in Sample 4-66 shows how a loop can be controlled using the *disable* statement.

Sample 4-66.
Loop control using the *disable* statement in Verilog

```
module loop_control;

initial
begin
    ...
    begin: exit_label
        while (...) begin: next_label
            ...
            // Force a new iteration
            if (...) disable next_label;

            ...
            // Break out of the loop
            if (...) disable exit_label;
            ...
        end // next_label
    end // exit_label
    ...
end

endmodule
```

Sample 4-67.
Non-blocking assignments potentially affected by a *disable* statement

```
module cpuif(...);

always
begin: if_logic
    ...
    data  <= #(Ta) read_val;
    dtack <= #(Tack) 1'b1;
    @ (negedge ale);
    data  <= #(Thold) 32'bz;
    dtack <= #(Thold) 1'b0;
    ...
end

always wait (reset == 1'b1)
begin
    disable if_logic;
    wait (reset != 1'b1);
end
endmodule
```

Non-blocking assignment values may be affected by the *disable* statement.

The Verilog standard does not specify what happens to still-pending values that were scheduled using a non-blocking assignment within a block that is disabled. Consider the code in Sample 4-67. When a reset condition is detected, the *always* block modeling the CPU interface is disabled to restart it from the beginning. What should happen to the various values assigned to the CPU interface signals *data* and *dtack* using non-blocking assignments, but that may not have been assigned to the registers yet? Depending on the simulator you are using, these values may be removed from the scheduled value queue and never make it to the intended registers, or they may remain unaffected by the *disable* statement. Avoid disabling a block where non-blocking assignments are performed.

Output Arguments on Disabled Tasks

Output values may not make it out of disabled tasks.

Another area where the behavior of Verilog is left unspecified is the value of output arguments in disabled tasks. Look at the code in Sample 4-68. The *read* task has an output argument returning the value that was read. Within the task, a *disable* statement is used to abort its execution at the end of the read cycle. Because the entire task was disabled, whether the value of *rdat* is copied out into the register *actual* used to invoke the task is not specified in the Verilog standard.

Sample 4-68.
Unspecified behavior of disabled tasks

```
task read;
   input  [7:0] radd;
   output [7:0] rdat;
begin
   . . .
   if (valid) begin
      rdat = data;
      disable read;
   end
   . . .
end
endtask

initial
begin: test_procedure
   reg [7:0] actual;

   read(8'hF0, actual);
   . . .
end
```

Disable the inner block instead of the task.

In some simulators, the value of *actual* is updated with the value of *rdat*, effectively completing the read cycle. In some others, the value of *actual* remains unchanged, leaving the read cycle incomplete. This unspecified behavior can be easily avoided by disabling the internal *begin/end* block inside the task instead of the task itself, as shown in Sample 4-69.

Sample 4-69.
Avoiding
unspecified
behavior of
disabled tasks

```
task read;
    input  [7:0] radd;
    output [7:0] rdat;
begin: read_cycle
    ...
    if (valid) begin
        rdat = data;
        disable read_cycle;
    end
    ...
end
endtask
```

Non-Reentrant Tasks

This is not an unspecified behavior.

Non-reentrant tasks are not really an unspecified behavior in Verilog. All simulators have non-reentrant tasks because every declaration in a Verilog model is static. No declaration is dynamically allocated upon invocation of a subprogram or entry into a block of code. I decided to include it in this section because it is a relatively common source of problems in Verilog models.

The same memory space is used for all invocations of a task.

When you declare a task or a function, the memory space for its arguments and all other locally declared registers is allocated at compile time. There is a single location for the subprogram. The memory is not allocated at runtime each time the task or function is invoked. Every time a subprogram is invoked, the <u>same</u> memory space is used. This does not cause problems in functions or in tasks that do not include @, #, or *wait* statements. The local data space is used in a single invocation. The memory space is no longer in use by the time a second invocation is made. However, if a task contains delay control statements, it may still be active when a second invocation is made.

A second invoca-
tion clobbers the
data space of an
active prior invo-
cation.

Examine the code in Sample 4-70. The task named *write* contains
delay control statements and is invoked from two different *initial*
blocks. In Figure 4-14(a), the content of the arguments, local to the
task, is shown after the invocation from the first *initial* block. While
this first invocation is waiting, the second *initial* block is executed
and invokes the *write* task again, setting its local arguments to the
values shown in Figure 4-14(b). When the first invocation resumes,
it continues its execution, using the arguments provided by the second invocation: its data space was overwritten. It goes on to write
the value 8'h34 at address 8'h5A.

Sample 4-70.
Non-reentrant
task

```
task write;
    input [7:0] wadd;
    input [7:0] wdat;
begin
    ad_dt <= wadd;
    ale   <= 1'b1;
    rw    <= 1'b1;
    @ (posedge rdy);
    ad_dt <= wdat;
    ale   <= 1'b0;
    @ (negedge rdy);
end
endtask

initial write(8'h5A, 8'h00);
initial write(8'hAD, 8'h34);
```

Figure 4-14.
Task data
space

wadd	8'h5A	wadd	8'hAD
wdat	8'h00	wdat	8'h34
	(a)		(b)

Concurrent task
activations may
not be so obvious.

The concurrent invocation of the same task in Sample 4-70 is pretty
obvious. But most of the time, the conditions where a task is activated more than once are much more obscure. In a large verification environment, with numerous tasks invoked under a complex
control structure, it is very easy to concurrently activate a task and
corrupt an entire testcase without you, or Verilog, being aware of it.

Use a semaphore
to detect concur-
rent task activa-
tion.

The best approach to avoid this fatal condition is to use a semaphore to detect concurrent activation, as shown in Sample 4-71.
The *in_use* register indicates whether or not the task is currently
activated. If the task is invoked while the *in_use* register is set to

'1', a concurrent invocation of the task is detected and the simulation is terminated. Since the data space of the task was already corrupted, there is no possibility of recovering from this error. Terminating the simulation is the only option. The problem must be fixed by retiming the access to the task to ensure that no concurrent invocation takes place.

Sample 4-71.
Guarding non-reentrant task using a semaphore

```
task write;
    input [7:0] wadd;
    input [7:0] wdat;

    reg in_use;
begin
    if (in_use === 1'b1) $stop;
    in_use = 1'b1;

    ad_dt <= wadd;
    ale   <= 1'b1;
    rw    <= 1'b1;
    @ (posedge rdy);
    ad_dt <= wdat;
    ale   <= 1'b0;
    @ (negedge rdy);

    in_use = 1'b0;
end
endtask
```

Invest in guarding tasks with delay control statements.

Personally, I put a guard around any task I write that contains delay control statements. This lets my model tell me immediately if I misused it. I can immediately fix the problem, without having to diagnose a testbench failure back to a concurrent task activation. The time invested in adding the simple semaphore is well worth it. If the task I write is to be used by others, the message produced by the concurrent activation detection specifically states that the error is not in my task code, but in <u>their</u> use of it. This has saved me many technical support calls.

SUMMARY

This chapter first presented the difference in approaching a synthesizable model compared to a behavioral model. Behavioral modelling requires a greater degree of discipline because of the greater freedom. Powerful encapsulation techniques allow a behavioral model to be structured to minimize maintenance. High-level data abstractions simplify the writing of a model by creating data types that are more natural to work with for a given transformation function.

This chapter also described how VHDL and Verilog are unique from traditional programming languages and how they are simulated. The simulation process introduces some peculiarities that a good behavioral modeler must be aware of. Some of these peculiarities can make Verilog models non-portables and have been described in details.

## CHAPTER 5

STIMULUS AND RESPONSE

The purpose of writing testbenches is to apply stimulus signals to a design and observe the response. That response must then be compared against the expected behavior.

This chapter shows how to apply stimulus and observe response.

In this chapter, I show how to generate the stimulus signals. The greatest challenge with stimulus is making sure they are an accurate representation of the environment, not just a simple case. In this chapter I also show how to observe response, and more importantly, how to compare it against expected values. The final part of this chapter covers techniques for communicating the predicted the output to the monitors.

The next chapter shows how to structure a test-bench.

In the next chapter, I show how to best structure the stimulus generators and response monitors and the testcases that use them to minimize maintenance, and increase reusability across testbenches. If you prefer a top-down perspective, I recommend you start with the next chapter then come back to this one.

SIMPLE STIMULUS

In this section, I explain how to generate deterministic waveforms. Various techniques are developed to best generate stimulus signals. I show how synchronized waveforms can be properly generated and how to avoid underconstraining stimulus. I also demonstrate how to encapsulate and package signal generation operations using bus-functional models.

Generating stimulus is the process of providing input signals to the design under verification as shown in Figure 5-1. From the perspective of the stimulus generator, every input of the design is an output of the generator.

Figure 5-1.
Stimulus
generation

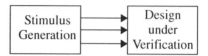

Generating a Simple Waveform

Clock signals are simple signals, but must be generated with care.

Because a clock signal has a very simple repetitive pattern, it is one of the first and most fundamental signals to generate. It is also the most critical signal to generate accurately. Many other signals use the clock signal to synchronize themselves.

The behavioral code to generate a 50 percent duty-cycle 100MHz clock signal is shown in Sample 5-1. To produce a more robust clock generator, use explicit assignments of values '0' and '1'. Using a statement like "clk = ~clk" would depend on the proper initialization of the clock signal to a value different than the default values of 1'bx or 'U'. Assigning explicit values also provides better control over the initial phase of the clock; you control whether the clock is starting high or low.

Sample 5-1.
Generating a
50% duty-
cycle clock

```
reg clk;
parameter cycle = 10; // 100MHz clock
always
begin
    #(cycle/2);
    clk = 1'b0;
    #(cycle/2);
    clk = 1'b1;
end
```

Any deterministic waveform is easy to generate.

Waveforms with deterministic edge-to-edge relationships with an easily identifiable period are also easy to generate. It is a simple process of generating each edge in sequence, at the appropriate time. For example, Figure 5-2 outlines an apparently complex waveform. However, Sample 5-2 shows that it is simple to generate.

Figure 5-2.
Apparently
complex
waveform

S   10 ns

Sample 5-2.
Generating a
deterministic
waveform

```
process
begin
    S <= '0'; wait for 20 ns;
    S <= '1'; wait for 10 ns;
    S <= '0'; wait for 10 ns;
    S <= '1'; wait for 20 ns;
    S <= '0'; wait for 50 ns;
    S <= '1'; wait for 10 ns;
    S <= '0'; wait for 20 ns;
    S <= '1'; wait for 10 ns;
    S <= '0'; wait for 20 ns;
    S <= '1'; wait for 40 ns;
    S <= '0'; wait for 20 ns;
    ...
end process;
```

The Verilog time-
scale may affect
the timing of
edges.

When generating waveforms in Verilog, you must select the appropriate timescale and precision to properly place the edges at the correct offset in time. When using an expression, such as "cycle/2", to compute delays, you must make sure that integer operations do not truncate a fractional part.

For example, the clock generated in Sample 5-3 produces a period of 14 ns because of truncation. If the precision is not sufficient, the delay values are rounded up or down, creating jitter on the edge location. For example, the clock generated in Sample 5-4 produces a period of 16 ns because of rounding. Only the signal generated in Sample 5-5 produces a 50 percent duty-cycle clock signal with a precise 15 ns period because the timescale offers the necessary precision for a 7.5 ns half-period.

Sample 5-3.
Truncation
errors in stim-
ulus genera-
tion

```
`timescale 1ns/1ns
module testbench;
...
reg clk;
parameter cycle = 15;
always
begin
   #(cycle/2);   // Integer division
   clk = 1'b0;
   #(cycle/2);   // Integer division
   clk = 1'b1;
end
endmodule
```

Sample 5-4.
Rounding
errors in stim-
ulus genera-
tion

```
`timescale 1ns/1ns
module testbench;
...
reg clk;
parameter cycle = 15;
always
begin
   #(cycle/2.0);    // Real division
   clk = 1'b0;
   #(cycle/2.0);    // Real division
   clk = 1'b1;
end
endmodule
```

Sample 5-5.
Proper preci-
sion in stimu-
lus generation

```
`timescale 1ns/100ps
module testbench;
...
reg clk;
parameter cycle = 15;
always
begin
   #(cycle/2.0);
   clk = 1'b0;
   #(cycle/2.0);
   clk = 1'b1;
end
endmodule
```

Generating a Complex Waveform

Avoid generating only a subset of possible complex waveforms.

A more complex waveform, with variations in the edge-to-edge timing relationships, requires more effort to model properly. Care must be taken not to overconstrain the waveform generation or to limit it to a subset of its possible variations. For example, if you generate the waveform illustrated in Figure 5-3 using the code in Sample 5-6, you generate only one of the many possible waveforms that meet the specification.

Figure 5-3.
Complex
waveform

	min	max
T_H	5	7
T_L	3	5

Sample 5-6.
Improperly
generating a
complex
waveform

```
process
begin
    S <= '0'; wait for (5 ns + 7 ns) / 2;
    S <= '1'; wait for (3 ns + 5 ns) / 2;
end process;
```

Use a random number generator to model uncertainty.

To properly generate the complex waveform as specified, it is necessary to model the uncertainty of the edge locations within the minimum and maximum delay range. This can be easily accomplished by randomly generating a delay within the valid range. Verilog has a built-in system task to generate 32-bit random values called *$random*. VHDL does not have a built-in random function, but public-domain packages of varying complexity are available.[1] The code in Sample 5-7 properly generates a non-deterministic complex waveform.

Sample 5-7.
Properly generating a complex waveform

```
process
begin
    S <= '0';
    wait for 5 ns + 2 ns * rnd_pkg.random;
    S <= '1';
    wait for 3 ns + 2 ns * rnd_pkg.random;
end process;
```

1. References to random number generation and linear-feedback shift register packages can be found in the *resources* section of:

   http://janick.bergeron.com/wtb

A linear random
distribution may
not yield enough
interesting values.

To generate waveforms that are likely to stress the design under
verification, it may be necessary to make sure that there are many
instances of absolute minimum and absolute maximum values.
With the linear random distribution produced by common random
number generators, this is almost impossible to guarantee. You
have to modify the waveform generator to issue more edges at the
extremes of the valid range than would otherwise be produced by a
purely linear random delay generation. In Sample 5-8, a function is
used to skew the random distribution with 30 percent minimum
value, 30 percent maximum value, and 40 percent random linear
distribution within the valid range.

Sample 5-8.
Skewing the
linear random
distribution

```
process
    function skewed_dist return real is
        variable distribute: real;
    begin
        distribute := rnd_pkg.random;
        if distribute < 0.3 then
            return 0.0;
        elsif distribute < 0.6 then
            return 1.0;
        else
            return rnd_pkg.random;
        end if;
    end skewed_dist;
begin
    S <= '0';
    wait for 5 ns + 2 ns * skewed_dist;
    S <= '1';
    wait for 3 ns + 2 ns * skewed_dist;
end process;
```

Generating Synchronized Waveforms

Most waveforms
are not indepen-
dent.

Stimuli for a design are never composed of a single signal. Multiple
signals must be properly generated with respect to each other to
properly stimulate the design under verification. When generating
interrelated waveforms, you must be careful not to create race conditions and to properly align edges both in real time and in delta
time.

Synchronized
waveforms must
be properly mod-
eled.

The first signal to be generated after the clock signal is the hardware reset signal. These two signals must be properly synchronized
to correctly reset the design. The generation of the reset signal
should also reflect its synchronization with the clock signal. For

example, consider the specification for a reset signal shown in Figure 5-4. The code in Sample 5-9 shows how such a waveform is frequently generated.

Figure 5-4.
Reset
waveform
specification

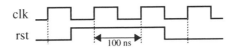

clk

rst

100 ns

Sample 5-9.
Improperly
generating a
synchronized
waveform

```
always
begin
    #50 clk = 1'b0;
    #50 clk = 1'b1;
end

initial
begin
    rst = 1'b0;
    #150 rst = 1'b1;
    #200 rst = 1'b0;
end
```

Race conditions can be easily created between synchronized signals.

There are two problems with the way these two waveforms are generated in Sample 5-9. The first problem is functional: there is a race condition between the *clk* and *rst* signals.[2] At simulation time 150, and again later at simulation time 350, both registers are assigned at the same timestep. Because the *blocking* assignment is used for both assignments, one of them is assigned first. A block sensitive to the falling edge of *clk* may execute before or after *rst* is assigned. From the perspective of that block, the specification shown in Figure 5-4 could appear to be violated. The race condition can be eliminated by using *non-blocking* assignments, as shown in Sample 5-10. Both *clk* and *rst* signals are assigned between timesteps, when no blocks are executing. If the design under verification uses the falling edge of *clk* as the active edge, *rst* is already - and reliably - assigned.

2. I did not bring up race conditions in the section titled "Read/Write Race Conditions" on page 141 just to conveniently forget about them here. Just to keep you on your toes, you'll see them appear throughout this book.

Sample 5-10.
Race-free generation of a synchronized waveform

```
always
begin
   #50 clk <= 1'b0;
   #50 clk <= 1'b1;
end

initial
begin
   rst = 1'b0;
   #150 rst <= 1'b1;
   #200 rst <= 1'b0;
end
```

Lack of maintainability can introduce functional errors.

The second problem, which is just as serious as the first one, is maintainability of the description. You could argue that the first problem is more serious, since it is functional. The entire simulation can produce the wrong result under certain conditions. Maintainability has no such functional impact. Or has it? What if you made a change as simple as changing the phase or frequency of the clock. How would you know to also change the generation of the reset signal to match the new clock waveform?

Conditions in real life are different than within the confines of this book.

In the context of Sample 5-10, with Figure 5-4 nearby, you would probably adjust the generation of the *rst* signal. But outside this book, in the real world, these two blocks could be separated by hundreds of lines, or even be in different files. The specification is usually a document one inch thick, printed on both sides. The timing diagram shown in Figure 5-4 could be buried in an anonymous appendix, while the pressing requirements of changing the clock frequency or phase was urgently stated in an email message. And you were busy debugging this other testbench when you received that pesky email message! Would you know to change the generation of the reset signal as well? I know I would not.

Model the synchronization within the generation.

Waiting for an apparently arbitrary delay can move out-of-sync with respect to the delay of the clock generation. A much better way of modeling synchronized waveforms is to include the synchronization in the generation of the dependent signals, as shown in Sample 5-11. The proper way to synchronize the *rst* signal with the *clk* signal is for the generator to wait for the significant synchronizing event, whenever it may occur. The timing or phase of the clock generator can now be modified, without affecting the proper generation of the *rst* waveform. From the perspective of a design, sensitive to the falling edge of *clk*, *rst* is reliably assigned one delta-cycle after the clock edge.

Sample 5-11.
Proper genera-
tion of a syn-
chronized
waveform

```
always
begin
    #50 clk <= 1'b0;
    #50 clk <= 1'b1;
end

initial
begin
    rst = 1'b0;
    wait (clk !== 1'bx);
    @ (negedge clk);
    rst <= 1'b1;
    @ (negedge clk);
    @ (negedge clk);
    rst <= 1'b0;
end
```

Synchronized
waveforms may be
generated from a
single block.

The maintainability argument can be taken one step further. Remember the section "Parallel vs. Sequential" on page 132? The sequence of the *rst* and *clk* waveforms is deterministic and can be modeled using a single sequential statement block, as shown in Sample 5-12. The synchronized portion of the *rst* and *clk* wave-forms is generated first, then the remaining free-running *clk* wave-form is generated. This generator differs from the one in Sample 5-11: from the perspective of a design sensitive to the falling edge of *clk*, *rst* has already been assigned.

Sample 5-12.
Sequential
generation of a
synchronized
waveform

```
initial
begin
    // Apply reset for first 2 clock cycles
    rst = 1'b0;
    #50 clk <= 1'b0;
    repeat (2) #50 clk <= ~clk;
    rst <= 1'b1;
    repeat (4) #50 clk <= ~clk;
    rst <= 1'b0;

    // Generate only the clock afterward
    forever #50 clk <= ~clk;
end
```

Delta delays are functionally equivalent to real delays.

In the specification shown in Figure 5-4, the transition of *rst* is aligned with a transition on *clk*. The various ways of generating these two signals determined the sequence of these transitions, whether they occurred at the same delta cycle, in different delta cycles, or if their ordering was deterministic. Although delta-cycle delays are considered zero-delays by the simulator, functionally they have the same effect as real delays.

The next two sections, "Aligning Waveforms in Delta-Time" and "Generating Synchronous Data Waveforms" discuss how signals can be properly aligned or delayed to prevent unintentional functional delays.

Aligning Waveforms in Delta-Time

Derived waveforms, such as the one shown in Figure 5-5, are apparently easy to generate. A simple process, sensitive to the proper edge of the original signal as shown in Sample 5-13, and voila! Even the waveform viewer shows that it is right!

Figure 5-5.
Derived
waveform
specification

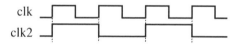

Sample 5-13.
Improperly
generating a
derived wave-
form

```
clk2_gen: process(clk)
begin
    if clk = '1' then
        clk2 <= not clk2;
    end if;
end process clk2_gen;
```

Watch for delta delays in derived waveforms.

The problem is not visually apparent. Because of the simulation cycle (See "The Simulation Cycle" on page 129), there is a delta cycle between the rising edge of the base clock signal, and the transition on the derived clock signal, as shown in Figure 5-6. Any data transferred from the base clock domain to the derived clock domain goes through this additional delta cycle delay. In a zero-delay simulation, such as a behavioral or RTL model, this additional delta-cycle delay can have the same effect as an entire clock cycle delay.

To maintain the relationship between the base and derived signals, their respective edges must be aligned in delta time. The only way to perform this task is to re-derive the base signal, as shown in Sample 5-14 and illustrated in Figure 5-7. The base signal is never used by other processes. Instead, they must use the derived base signal.

Figure 5-6.
Delta delay in derived waveform

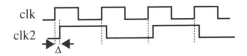

Sample 5-14.
Properly generating a derived waveform

```
derived_gen: process(clk)
begin
    clk1 <= clk;
    if clk = '1' then
        clk2 <= not clk2;
    end if;
end process derived_gen;
```

Figure 5-7.
Generation of aligned derived signals

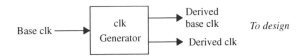

Generating Synchronous Data Waveforms

There is a race condition between the clock and data signal.

Sample 5-10, Sample 5-11, and Sample 5-15 show how you could generate a zero-delay synchronous data waveform. In Sample 5-11 and Sample 5-15, it is identical to the way flip-flops are inferred in an RTL model. As illustrated in Figure 5-8, there is a delay between the edge on the clock and the transition on *data,* but it is a single delta cycle. In terms of simulation time, there is no delay. For RTL models, this infinitesimal clock-to-Q delay is sufficient to properly model the behavior of synchronous circuits. However, this assumes that all clock edges are aligned in delta time (see "Aligning Waveforms in Delta-Time" on page 164). If you are generating both clock and data signals from the outside of the model of the design under verification, you have no way of ensuring that the total number of delta-cycle delays between the clock and the data is maintained, or at least be in favor of the data signal!

Sample 5-15.
Zero-delay
generation of
synchronous
data

```
sync_data_gen: process(clk)
begin
    if clk = '0' then
        data <= ...;
    end if;
end process sync_data_gen;
```

Figure 5-8.
Synchronous
data
waveforms

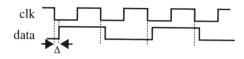

The clock may be
delayed more than
the data.

For many possible reasons, the clock signal may be delayed by more delta cycles than its corresponding data signal. These delays could be introduced by using different I/O pad models for the clock and data pins. They could also be introduced by the clock distribution network, which does not exist on the data signal. If the clock signal is delayed more than the data signal, even in zero-time as shown in Figure 5-9, the effect is the same as removing an entire clock cycle from the data path.

Interface specifications never specify zero-delay values. A physical interface always has a real delay between the active edge of a clock signal and its synchronous data. When generating synchronous data, always provide a real delay between the active edge and the transition on the data signal, as shown in Sample 5-16 and Sample 5-17.

Figure 5-9.
Delta delays in
clock path

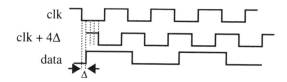

Sample 5-16.
Non-zero-
delay genera-
tion of syn-
chronous data

```
sync_data_gen: process(clk)
begin
    if clk = '0' then
        data <= ... after 1 ns;
    end if;
end process sync_data_gen;
```

Sample 5-17.
Sequential
generation of a
delayed syn-
chronized data
waveform
	``
	initial
	begin
	    // Apply reset for first 2 clock cycles
	    rst = 1'b0;
	    #50 clk <= 1'b0;
	    repeat (2) #50 clk <= ~clk;
	    rst <= #1 1'b1;
	    repeat (4) #50 clk <= ~clk;
	    rst <= #1 1'b0;
	
	    // Generate only the clock afterward
	    forever #50 clk <= ~clk;
	end
	``

Encapsulating Waveform Generation

The generation of waveforms may need to be repeated during a simulation.

There is a problem with the way the *rst* waveform is generated in Sample 5-17. What if it were necessary to reset the device under verification multiple times during the execution of a testbench? One example would be to execute multiple testcases in a single simulation. Another one is the "hardware reset" testcase which verifies that the reset operates properly. In that respect, the code in Sample 5-11 is closer to an appropriate solution. The only thing that needs to be changed is the use of the *initial* block. The *initial* block runs only once and is eliminated from the simulation once completed. There is no way to have it execute again during a simulation.

Encapsulate waveform generation in a subprogram.

The proper mechanism to encapsulate statements that you may need to repeat during a simulation is to use a *task* or a *procedure* as shown in Sample 5-18. To repeat the waveform, simply call the subprogram. To maintain the behavior of using an *initial* block to automatically reset the device under verification at the beginning of the simulation, simply call the task in an *initial* block. Pop quiz: what is missing from the *hw_reset* task in Sample 5-18? The answer can be found in this footnote.[3]

A subprogram can be used to properly apply vectors.

Another example of a synchronized waveform whose generation can be encapsulated is the application of input vector data. As illus-

---

3. The task *hw_reset* contains delay control statements. It should contain a semaphore to detect concurrent activation. You can read more about this issue in "Non-Reentrant Tasks" on page 151.

Sample 5-18. Encapsulating the generation of a synchronized waveform

```
always
begin
    #50 clk <= 1'b0;
    #50 clk <= 1'b1;
end

task hw_reset;
begin
    rst = 1'b0;
    wait (clk !== 1'bx);
    @ (negedge clk);
    rst <= 1'b1;
    @ (negedge clk);
    @ (negedge clk);
    rst <= 1'b0;
end
endtask
initial hw_reset;
```

trated in Figure 5-10, vector data must be applied with a proper setup and hold time - but no more - to meet the input timing constraints. Instead of repeating the synchronization for each vector, a subprogram can be used for synchronization with the input clock. It would also apply the vector data received as input argument. The code in Sample 5-19 shows the implementation and use of a *task* applying input vectors according to the specification in Figure 5-10. Notice how the input is set to unknowns after the specified hold time to stress the timing of the interface. Leaving the input to a constant value would not detect cases where the device under verification does not meet the maximum hold requirement.

Figure 5-10. Input data waveform specification

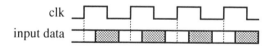

Sample 5-19.
Encapsulating
the applica-
tion of input
data vectors

```
task apply_vector;
    input [...] vector;
begin
    inputs <= vector;
    @(posedge clk);
    #(Thold);
    inputs <= ...'bx;
    #(cycle - Thold - Tsetup);
end
endtask

initial
begin
    hw_reset;
    apply_vector(...);
    apply_vector(...);
    ...
end
```

Abstracting Waveform Generation

Vectors are diffi-
cult to write and
maintain.

Using synchronous test vectors to verify a design is rather cumbersome. They are hard to interpret and difficult to correctly specify.
For example, using vectors to verify a synchronously resetable D
flip-flop with a 2-to-1 multiplexer on the input, as shown in
Figure 5-11, could be stimulated using the vectors shown in Sample
5-20.

Figure 5-11.
2-to-1 input
sync reset D
flip-flop.

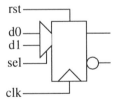

Use subprograms
to encapsulate
operations.

It would be easier if the operation accomplished by the vectors
were abstracted. The device under verification can only perform
three things:

- A synchronous reset

- Load from input *d0*

- Load from input *d1*

Sample 5-20.
Test vectors
for 2-to-1
input sync
reset D flip-
flop

```
initial
begin
    // Vector: rst, d0, d1, sel
    apply_vector(4'b1110);
    apply_vector(4'b0100);
    apply_vector(4'b1111);
    apply_vector(4'b0011);
    apply_vector(4'b0010);
    apply_vector(4'b0011);
    apply_vector(4'b1111);
    ...
end
```

Instead of providing vectors to repeatedly perform these operations, why not provide subprograms that perform these operations? All that will be left is to call the subprograms in the appropriate order, with the appropriate data.

Try to apply the worst possible combination of inputs.

The subprogram to perform the synchronous reset is very simple. It needs to assert the *rst* input, then wait for the active edge of the clock. But what about the other inputs? You could decide to leave them unchanged, but is that the worst possible case? What if the reset was not functional and the device loaded one of the inputs and that input was set to '0'? It would be impossible to differentiate the wrong behavior from the correct one. To create the worst possible condition, both *d0* and *d1* inputs must be set to '1'. The *sel* input can be set randomly, since either input selection should be functionally identical. An implementation of the *sync_reset* task is shown in Sample 5-21.

Sample 5-21.
Abstracting
the sync reset
operation

```
task sync_reset;
begin
    rst <= 1'b1;
    d0  <= 1'b1;
    d1  <= 1'b1;
    sel <= $random;
    @ (posedge clk);
    #(Thold);
    {rst, d0, d1, sel} <= 4'bxxxx;
    #(cycle - Thold - Tsetup);
end
endtask
```

Pass input values
as arguments to
the subprogram.

The second operation this design can perform is to load input *d0*. The *task* to perform this operation is shown in Sample 5-22. Unlike resetting the design, loading data can have different input values: it can load either a '1' or a '0'. The value of the input to load is passed as an argument to the task. The worst condition is created when the other input is set to the complement of the input value on *d0*. If the device is not functioning properly and is loading from the wrong input, then the result will be clearly wrong.

Sample 5-22.
Abstracting
the load *d0*
operation

```
task load_d0;
    input data;
begin
    rst <= 1'b0;
    d0  <= data;
    d1  <= ~data;
    sel <= 1'b0;
    @ (posedge clk);
    #(Thold);
    {rst, d0, d1, sel} <= 4'bxxxx;
    #(cycle - Thold - Tsetup);
end
endtask
```

Stimulus gener-
ated with
abstracted opera-
tions is easier to
write and main-
tain.

The last operation this design can perform is to load input *d1*. The *task* abstracting the operation to load from input *d1* is similar to the one shown in Sample 5-22. Once operation abstractions are available, providing the proper stimulus to the design under verification is easy to write and understand. Compare the code in Sample 5-23 with the code of Sample 5-20. If the polarity of the *rst* input were changed, which verification approach would be easiest to modify?

Sample 5-23.
Verifying the
design using
operation
abstractions

```
initial
begin
    sync_reset;
    load_d0(1'b1);
    sync_reset;
    load_d1(1'b1);
    load_d0(1'b0);
    load_d1(1'b1);
    sync_reset;
    . . .
end
```

VERIFYING THE OUTPUT

Generating stimulus is only half of the job. Actually, it is more like 30 percent of the job. The other part, verifying that the output is as expected, is much more time-consuming and error-prone. There are various ways the output can be checked against expectations. They have varying degrees of applicability and repeatability.

Visual Inspection of Response

Results can be printed.

The most obvious method for verifying the output of a simulation is to visually inspect the results. The visual display can be an ASCII printout of the input and output values at specific points in time, as shown in Sample 5-24.

Sample 5-24. ASCII view of simulation results	r s sddeqq Time t011 b ----------
	0100 1110xx
	0105 111001
	0200 010001
	0205 010010
	0300 111110
	0305 111101
	0400 001101
	0405 001110
	0500 001010
	0505 001010
	0600 001110
	0605 001110
	0700 111110
	0705 111101
	. . .

Producing Simulation Results

To print simulation results, you must model the signal sampling.

The specific points in time that are significant for a particular design or testbench are always different. Which signals are significant is also different and may change as the simulation progresses. If you know which time points and signals are significant for determining the correctness of the simulation results, you have to be able to model that knowledge. Producing the proper simulation results involves modeling the behavior of the signal sampling.

Many sampling techniques can be used.	There are many sampling techniques, each as valid as the other. The correct sampling technique depends on your needs and on what makes the simulation results significant. Just as you have to decide which input sequence is relevant for the functionality you are trying to verify, you must also decide on the output sampling that is relevant for determining the success or failure of the function under verification.
You can sample at regular intervals.	The simplest sampling technique is to sample the relevant signals at a regular interval. The interval can be an absolute delay value, as illustrated in Sample 5-25, or a reference signal such as the clock, as illustrated in Sample 5-26.

Sample 5-25.
Sampling at a
delay interval

```
parameter INTERVAL = 10;
always
begin
    #(INTERVAL);
    $write(...);
end
```

Sample 5-26.
Sampling
based on a ref-
erence signal

```
process (clk)
    variable L: line;
begin
    if clk'event and clk = '0' then
        write(L, ...);
        writeline(output, L);
    end if;
end process;
```

You can sample based on a signal changing value.	Another popular sampling technique is to sample a set of signals whenever one of them changes. This is used to reduce the amount of data produced during a simulation when signals do not change at a constant interval.

To sample a set of signals, simply make a *process* or *always* block sensitive to the signals whose changes are significant, as shown in Sample 5-27. The set of signals displayed and monitored can be different. Verilog has a built-in task, called *$monitor*, to perform this sampling when the set of display and monitored signals are identical.

An example of using the *$monitor* task is shown in Sample 5-28. Its behavior is different from the VHDL sampling process shown in Sample 5-27: changes in values of signals *rst*, *d0*, *d1*, *sel*, *q*, and *qb* cause the display of simulation results, whereas only changes in *q* and *qb* trigger the sampling in the VHDL example. Note that Verilog simulations are limited to a single active *$monitor* task. Any subsequent call to *$monitor* <u>replaces</u> the previous monitor.

Sample 5-27.
Sampling based on signal changes

```
process (q, qb)
    variable L: line;
begin
    write(L, rst & d0 & d1 & sel & q & qb);
    writeline(output, L);
end process;
```

Sample 5-28.
Sampling using the *$monitor* task

```
initial
begin
    $monitor("...", rst, d0, d1, sel, q, qb);
end
```

Minimizing Sampling

To improve simulation performance, minimize sampling.

The use of an output device on a computer slows down the execution of any program. Therefore, the production of simulation output reduces the performance of the simulation. To maximize the speed of a simulation, minimize the amount of simulation output produced during its execution.

In Verilog, an active *$monitor* task can be turned on and off by using the *$monitoron* and *$monitoroff* tasks, respectively. If you are using an explicit sampling *always* block or are using VHDL, you should include sampling minimization techniques in your model, as illustrated in Sample 5-29. A very efficient way of minimizing sampling is to have the stimulus turn on the sampling when an interesting section of the testcase is entered, as shown in Sample 5-30.

Visual Inspection of Waveforms

Results are better viewed when plotted over time.

Waveform displays usually provide a more intuitive visual representation of simulation results. Figure 5-12 shows the same information as Sample 5-24, but using a waveform view. The waveform view has the advantage of providing a continuous display of many values over the entire simulation time, not just at specific time points as in a text view. Therefore, you need not specify or model a particular sampling technique. The signals are continuously sam-

Sample 5-29.
Minimizing
sampling

```
process
begin
   wait until <interesting_condition>;
   sampling: loop
       wait on q, qb;
       write(l, rst & d0 & d1 & sel & q & qb);
       writeline(output, l);
       exit sampling
           when not <interesting_condition>;
   end loop sampling;
end process;
```

Sample 5-30.
Controlling
the sampling
from the stim-
ulus

```
initial
begin
    $monitor("...", rst, d0, d1, sel, q, qb);
    $monitoroff;
    sync_reset;
    load_d0(1'b1);
    sync_reset;
    $monitoron;
    load_d1(1'b1);
    load_d0(1'b0);
    load_d1(1'b1);
    sync_reset;
    $monitoroff;

    ...
end
```

pled, usually into an efficient database format. Sampling for waveforms must be turned on explicitly. It is a tool-dependent process
that is different for each language and each tool.

Figure 5-12.
Waveform
view of
simulation
results

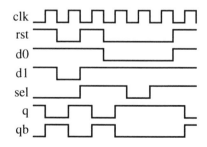

Minimize the
number and dura-
tion of sampled
signals.

The default behavior is to sample all signals during the entire simulation. The waveform sampling process consumes a significant portion of the simulation resources. Reducing the number of signals

sampled, or the duration of the sampling, increases the simulation performance.

SELF-CHECKING TESTBENCHES

This section introduces a reliable and reproduceable technique for output verification: testbenches that verify themselves. I discuss the pros and cons of popular vector-based implementation techniques. I show how to verify the simulation results at run-time by modelling the expected response at the same time as the stimulus.

Visual inspection is not acceptable.

The model of the D flip-flop with a 2-to-1 input mux being verified has a functional error. Can you identify it using either views of the simulation results in Sample 5-24 or Figure 5-12? How long did it take to diagnose the problem?[4]

This example was for a very simple design, over a very short period of time, and for a very small number of signals. Imagine visually inspecting simulation results spanning hundreds of thousands of clock cycles, and involving hundreds of input and output signals. Then imagine repeating this visual inspection for every testbench, and every simulation of every testbench. The probability that you will miss identifying an error is equal to one. You must automate the process of comparing the simulation results against the expected outputs.

Input and Output Vectors

Specify the expected output values for each clock cycle.

The first step in automating output verification is to include the expected output with the input stimulus for every clock cycle. The vector application task in Sample 5-19 can be easily modified to include the comparison of the output signals with the specified output vector, as shown in Sample 5-31. The testcase becomes a series of input/output test vectors, as shown in Sample 5-32.

Test vectors require synchronous interfaces.

The main problem with input and output test vectors (other than the fact that they are very difficult to specify, maintain, and debug), is that they require perfectly synchronous interfaces. If the design under verification contains interfaces in different clock domains,

---

4. The logic value on input *d0* is ignored and a '1' is always loaded.

Sample 5-31.
Application of input and verification of output data vectors

```
task apply_vector;
    input [...] in_data;
    input [...] out_data;
begin
    inputs <= in_data;
    @(posedge clk);
    fork
        begin
            #(Thold);
            inputs <= ...'bx;
        end
        begin
            #(Td);
            if (outputs !== out_data) ...;
        end
        #(cycle - Thold - Tsetup);
    join
end
endtask
```

Sample 5-32.
Input/output test vectors for 2-to-1 input sync reset D flip-flop

```
initial
begin
    // In: rst, d0, d1, sel
    // Out: q, qb
    apply_vector(4'b1110, 2'b00);
    apply_vector(4'b0100, 2'b10);
    apply_vector(4'b1111, 2'b00);
    apply_vector(4'b0011, 2'b10);
    apply_vector(4'b0010, 2'b01);
    apply_vector(4'b0011, 2'b10);
    apply_vector(4'b1111, 2'b00);
    ...
end
```

each requires its own test vector stream. If any interface contains asynchronous signals, they have to be either externally synchronized before vectors are applied, or treated as synchronous signals, therefore under-constraining the verification.

Golden Vectors

A set of reference simulation results can be used.

The next step toward automation of the output verification is the use of golden vectors. It is a simple extension of the manufacturing test process where devices are physically subjected to a series of qualifying test vectors. A set of reference output results, determined to be correct, are kept in a file or database. The simulation outputs are captured in a similar format during a simulation. They are then compared against the reference results. Golden vectors have an advantage over input/output vectors because the expected output values need not be specified in advance.

Text files can be compared using *diff*.

If the simulation results are kept in ASCII files, the simplest comparison process involves using the UNIX *diff* utility. The *diff* output for the simulation results shown in Sample 5-24 is shown in Sample 5-33. You can appreciate how difficult the subsequent task of diagnosing the functional error will be.

Sample 5-33. *diff* output of comparing ASCII view of simulation results

```
14c2
>0505 001010
>0600 001110
--------
<0505 001001
<0600 001110
. . .
```

Waveforms can be compared by a specialized tool.

Waveform comparators can also be used. They are tools similar to waveform viewers and are usually built into one. They compare two sets of waveforms then highlight the differences on a graphical display. The display of a waveform comparator might look something like the results illustrated in Figure 5-13. Identifying the problem is easier since you have access to the entire history of the simulation in a single view.

Figure 5-13. Waveform differences in simulation results

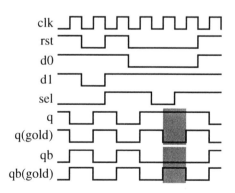

Writing Testbenches: Functional Verification of HDL Models

Golden vectors must still be visually inspected.	The main problem with golden simulation results is that they need to be visually inspected to be determined as valid. This self-checking technique only reduces the number of times a set of simulation responses must be visually verified, not the need for visual inspection. The result from each testbench must still be manually confirmed as good.
Golden vectors do not adapt to changes.	Another problem: reference simulation results do not adapt to modifications in the design under verification that may only affect the timing of the result, without affecting its functional correctness. For example, an extra register may be added in the datapath of a design to help meet timing constraints. All that was added was a pipeline delay. The functionality was not modified. Only the latency was increased. If that latency is irrelevant to the functional correctness of the overall system, the reference vectors must be updated to reflect that change.
Golden vectors require a significant maintenance effort.	Reference simulation results must be visually inspected for every testcase, and modified or regenerated whenever a change is made to the design, each time requiring visual inspection. Using reference vectors is a high-maintenance, low-efficiency self-checking strategy. Verification vectors should be used only when a design must be 100 percent backward compatible with an existing device, signal for signal, clock cycle for clock cycle. In those circumstances, the reference vectors never change and never require visual inspection as they are golden by definition.
Separate the reference vectors along clock domains.	Reference simulation results also work best with synchronous interfaces. If you have multiple interfaces in separate clock domains, it is necessary to generate reference results for each domain in a separate file. If a single file is used, the asynchronous relationship between the clock domains may result in the samples from different domains being written in a different order. The ordering difference is not functionally relevant, but would be flagged as an error by the comparison tool.

Run-Time Result Verification

Comparing simulation results against a reference set of vectors or waveforms is a post-processing process. It is possible to verify the correctness of the simulation results at runtime, in parallel with the stimulus generation.

You can use a reference model.

Using a reference model is a simple extension of the golden vector technique. As illustrated in Figure 5-14, the reference model and the design under verification are subjected to the same stimulus and their output is constantly monitored and compared for discrepancies.

Figure 5-14. Using a reference model

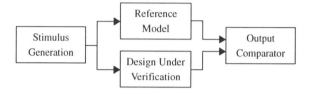

In reality, a reference model never works.

The reality of reference models is different. They rarely exist. When they do, they are in a language that cannot be easily integrated with either VHDL or Verilog. When they can be integrated, they produce output with a different timing or accuracy, making the output comparison impractical. When all of these obstacles are overcome, they are often a burden on the simulation performance. Using reference simulation results, as described in the previous section, is probably a better alternative.

You can model the expected response.

If you know what you are looking for when visually inspecting simulation results, you should be able to describe it also. It should be part of the testcase specification. If the expected response can be described, it can be modeled. If it can be modeled, it can be included in the testbench. By including the expected response in the testbench, it is able to determine automatically whether the testcase succeeded or failed.

Focus on operations instead of input and output vectors.

In "Abstracting Waveform Generation" on page 169, subprograms were used to apply stimulus to the design. These subprograms abstracted the vectors into atomic operations that could be performed on the design. Why not include the verification of the operation's output as part of the subprogram? Instead of simply applying inputs, then leaving the output verification to a separate process, integrate both the stimulus and response checking into complete operations. Performing the verification becomes a matter of verifying that operations, individually and in sequence, are performed appropriately.

For example, the task shown in Sample 5-21 can include the verification that the flip-flop was properly reset as shown in Sample 5-

34. Similarly, the task used to apply the stimulus to load data from the *d0* input shown in Sample 5-22 can be modified to include the verification of the output, as shown in Sample 5-35. The testcase shown in Sample 5-23 now becomes entirely self-checking.

Sample 5-34.
Verifying the
sync reset
operation

```
task sync_reset;
begin
    rst <= 1'b1;
    d0  <= 1'b1;
    d1  <= 1'b1;
    sel <= $random;
    @ (posedge clk);
    #(Thold);
    if (q !== 1'b0 || qb !== 1'b1) ...
    {rst, d0, d1, sel} <= 4'bxxxx;
    #(cycle - Thold - Tsetup);
end
endtask
```

Sample 5-35.
Verifying the
load *d0* opera-
tion

```
task load_d0;
    input data;
begin
    rst <= 1'b0;
    d0  <= data;
    d1  <= ~data;
    sel <= 1'b0;
    @ (posedge clk);
    #(Thold);
    if (q !== data || qb !== ~data) ...
    {rst, d0, d1, sel} <= 4'bxxxx;
    #(cycle - Thold - Tsetup);
end
endtask
```

Make sure the out-
put is properly ver-
ified.

The problem with output verification is that you can't identify a functional discrepancy if you are not looking at it. Using an *if* statement to verify the output in the middle of a stimulus process only looks at the output value for a brief instant. That may be acceptable, but it does not say anything about the *stability* of that output. For example, the tasks in Sample 5-34 and Sample 5-35 only check the value of the output at a single point. Figure 5-15 shows the complete specification for the flip-flop. The verification sampling point is shown as well.

Figure 5-15.
Timing
specification
for the flip-
flop

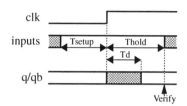

Make sure you
verify the output
over the entire sig-
nificant time
period.

To properly and completely verify the functionality of the design, it is necessary to verify that the output is stable, except for the short period after the rising edge of the clock. That could be easily verified using a static timing analysis tool and a set of suitable constraints to verify against. If you want to perform the verification in Verilog or VHDL, the stability of the output cannot be easily verified in the same subprogram that applies the input. The input follows a deterministic data and timing sequence, whereas monitoring stability requires that the testbench code be ready to react to any unexpected changes. Instead, it is better to use a separate monitor process, executing in parallel with the stimulus. The stimulus subprogram can still check the value. The stability monitor, as shown in Sample 5-36, simply verifies that the output remains stable, whatever its value. In VHDL, the *'stable* attribute was designed for this type of application, as shown in Sample 5-37. The stability of the output signal can be verified in the stimulus procedure, but it requires prior knowledge of the clock period to perform the timing check.

Sample 5-36.
Verifying the
stability of
flip-flop out-
puts

```
initial
begin
    // wait for the first clock edge
    @ (posedge clk);
    forever begin
        // Ignore changes for Td after clock edge
        #(Td);
        // Watch for a change before the next clk
        fork: stability_mon
            @ (q or qb) $write("...");
            @ (posedge clk) disable stability_mon;
        join
    end
end
```

Writing Testbenches: Functional Verification of HDL Models

<table>
<tr>
<td>

Sample 5-37.
Verifying the
load *d0* opera-
tion and out-
put stability

</td>
<td>

```
procedure load_d0(data: std_logic) is
begin
  rst <= '0';
  d0  <= data;
  d1  <= not data;
  sel <= '0';
  wait until clk = '1';
  assert q'stable(cycle - Td) and
         qb'stable(cycle - Td);
  wait for Thold;
  rst <= 'X';
  d0  <= 'X';
  d1  <= 'X';
  sel <= 'X';
  assert q = data and qb = not data;
  wait for cycle - Thold - Tsetup;
end load_d0;
```

</td>
</tr>
</table>

COMPLEX STIMULUS

This section introduces more complex stimulus generation scenarios through the use of bus-functional models. I start with non-deterministic stimulus, where the stimulus or its timing depends on answers from the device under verification. I also show how to avoid wasting precious simulation cycles by getting caught in deadlock conditions. I explain how to generate asynchronous stimulus and more complex protocols such as CPU cycles. Finally, I show how to write configurable bus-functional models.

Generating inputs may require cooperating with the design.

Applying stimulus to a clock or reset input is straightforward. You are under complete control of the timing of the input signal. However, if the interface being driven contains handshaking or flow-control signals, the generation of the stimulus requires cooperation with the design under verification.

Feedback Between Stimulus and Design

Without feedback, verification can be under-constrained.

Figure 5-16 shows the specification for a simple bus arbiter. If you were to verify the design of the arbiter using test vectors applied at every clock cycle, as described in "Input and Output Vectors" on page 176, you would have to assume a specific delay between the assertion of the *req* signal and the assertion of the *grt* signal. Any delay value between one and five clock cycles would be functionally correct, but the only reliable choice is a delay of five cycles.

Similarly, a delay of three clock cycles would have to be made for the release portion of the verification. These choices, however, severely under-constrain the verification. If you want to stress the arbiter by issuing requests as fast as possible, you would want to know when the request was granted and released, so it could be reapplied as quickly as possible.

Figure 5-16.
Specification
for a simple
arbiter

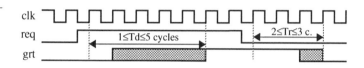

Stimulus genera-
tion can wait for
feedback before
proceeding.

If, instead of using input and output test vectors, you are using encapsulated operations to verify the design, you can modify the operation to wait for feedback from the design under verification before proceeding. You should also include any timing and functional verification in the feedback monitoring to ensure that the design responds in an appropriate manner. Sample 5-38 shows the *bus_request* operation procedure. It samples the *grt* signal at every clock cycle, and immediately returns once it detects that the bus was granted. With a similarly implemented *bus_release* procedure, a testcase that stresses the arbiter under maximum load can be easily written, as shown in Sample 5-39.

Sample 5-38.
Verifying the
bus request
operation

```
procedure bus_request is
   variable cycle_count: integer := 0;
begin
   req <= '1';
   wait until clk = '1';
   while grt = '0' loop
      wait until clk = '1';
      cycle_count := cycle_count + 1;
   end loop;
   assert 1 <= cycle_count and cycle_count <= 5;
end bus_request;
```

Recovering from Deadlocks

A deadlock may
prevent the
testcase from run-
ning to comple-
tion.

There is a risk inherent to using feedback in generating stimulus: the stimulus now depends on the proper operation of the design under verification to complete. If the design does not provide the feedback as expected, the stimulus generation may be halted, waiting for a condition that will never occur. For example, consider the

Sample 5-39.
Stressing the
bus arbiter.

```
test_sequence: process
   procedure bus_request ...
   procedure bus_release ...
begin
   for I in 1 to 10 loop
      bus_request;
      bus_release;
   end loop;
   assert false severity failure;
end process test_sequence;
```

bus_request procedure in Sample 5-38. What happens if the *grt* signal is never asserted? The procedure remains stuck in the *while* loop and never returns.

A deadlocked simulation appears to be running correctly.

If this were to occur, the simulation would still be running, merrily going around and around the *while* loop. The simulation time would advance at each tick of the clock. The CPU usage of your workstation would show near 100 percent usage. The only symptom that something is wrong would be that no messages are produced on the simulation's output log and the simulation runs for much longer than usual. If you are watching the simulation run and expect regular messages to be produced during its execution, you would quickly recognize that something is wrong and manually interrupt it.

A deadlocked simulation wastes regression runs.

But what if there is no one watching the simulation, such as during a regression run? Regressions are large scale simulation runs where all available testcases are executed. They are used to verify that the functionality of the design under verification is still correct after modifications. Because of the large number of testcases involved in a regression, the process is automated to run unattended, usually overnight and on many computers. If a design modification creates a deadlock situation, all testcases scheduled to execute subsequently will never run, as the deadlocked testcase never terminates. The opportunity of detecting other problems in the regression run is wasted. It will be necessary to wait for another 24-hour period before knowing if the new version of the design meets its functional specification.

Eliminate the possibility of deadlock conditions.

When generating stimulus, you must make sure that there is no possibility of a deadlock condition. You must assume that the feedback condition you are waiting for may never occur. If the feedback condition fails to happen, you must then take appropriate action. It could include terminating the testcase, or jumping to the next portion of the testcase that does not depend on the current operation, or attempting to repeat the operation after some delay. Sample 5-38 was modified as shown in Sample 5-40 to avoid the deadlock condition created if the arbiter failed and the *grt* signal was never asserted.

Sample 5-40.
Avoiding
deadlock in
the bus request
operation

```
procedure bus_request is
    variable cycle_count: integer := 0;
begin
    req <= '1';
    wait until clk = '1';
    while grt = '0' loop
        wait until clk = '1';
        cycle_count := cycle_count + 1;
        assert cycle_count < 500
            report "Arbiter is not working"
            severity failure;
    end loop;
    assert 1 <= cycle_count and cycle_count <= 5;
end bus_request;
```

Sample 5-41.
Returning sta-
tus in the bus
request opera-
tion

```
procedure bus_request(good: out boolean) is
    variable cycle_count: integer := 0;
begin
    good := true;
    req <= '1';
    wait until clk = '1';
    while grt = '0' loop
        wait until clk = '1';
        cycle_count := cycle_count + 1;
        if cycle_count > 500 then
            good := false;
            return;
        end if;
    end loop;
    assert 1 <= cycle_count and cycle_count <= 5;
end bus_request;
```

Operation subpro-
grams could return
status.

If a failure of the feedback condition is detected, terminating the simulation on the spot, as shown in Sample 5-40, is easy to implement in each operation subprogram. If you want more flexibility in handling a non-fatal error, you might want to let the testcase handle the error recovery, instead of handling it inside the operation subprogram. The subprogram must provide an indication of the status of the operation's completion back to the testcase. Sample 5-41 shows the *bus_request* procedure that includes a *good* status flag indicating whether the bus was granted or not. The testcase is then free to attempt other bus request operations until it succeeds, as shown in Sample 5-42. Notice how the testcase takes care of avoiding its own deadlock condition if the bus request operation never succeeds.

Sample 5-42.
Handling failures in the
bus_request
procedure

```
testcase: process
   variable granted : boolean;
   variable attempts: integer := 0;
begin
   ...
   attempts := 0;
   loop
      bus_request(granted);
      exit when granted;
      attempts := attempts + 1;
      assert attempts < 5
         report "Bus was never granted"
         severity failure;
   end loop;
   ...
end process testcase;
```

Asynchronous Interfaces

Test vectors under-constrain asynchronous interfaces.

Test vectors are inherently synchronous. The inputs are all applied at the same time. The outputs are all verified at the same time. And this process is repeated at regular intervals. Many interfaces, although implemented using finite state machines and edge-triggered flip-flops, are <u>specified</u> in an asynchronous fashion. The implementer has arbitrarily chosen a clock to streamline the physical implementation of the interface. If that clock is not part of the specification, it should not be part of the verification. For example, Figure 5-17 shows an asynchronous specification for a bus arbiter. Given a suitable clock frequency, the synchronous specification shown in Figure 5-16 would be a suitable implementation.

Figure 5-17.
Asynchronous
specification
for a simple
arbiter

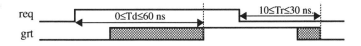

Verify the synchronous implementation against the asynchronous specification.

Even though a clock may be present in the implementation, if it is not part of the specification, you cannot use it to generate stimulus nor to verify the response. You would be verifying against a particular implementation, not the specification. For example, a VME bus is asynchronous. The verification of a VME interface cannot make use of the clock used to implement that interface. If a clock is present, and the timing constraints make reference to clock edges, then you <u>must</u> use it to generate stimulus and verify response. For example, a PCI bus is synchronous. A verification of a PCI interface must use the PCI system clock to verify any implementation.

Behavioral code does not require a clock like RTL code.

Testbenches are written using behavioral code. Behavioral models do not require a clock. Clocks are artifices of the implementation methodology and are required only for RTL code. The bus request phase of the asynchronous interface specified in Figure 5-17 can be verified asynchronously with the *bus_request* procedure shown in Sample 5-43 or Sample 5-44. Notice how neither model of the bus request operation uses a clock for timing control. Also, notice how the Verilog version, in Sample 5-44, uses the definitely non-synthesizeable *fork/join* statement to wait for the rising edge of *grt* for a maximum of 60 time units.

Sample 5-43.
Verifying the asynchronous bus request operation in VHDL

```
procedure bus_request(good: out boolean) is
begin
    req <= '1';
    wait until grt = '1' for 60 ns;
    good := grt = '1';
end bus_request;
```

Sample 5-44.
Verifying the asynchronous bus request operation in Verilog

```
task bus_request;
    output good;
begin
    req = 1'b1;
    fork: wait_for_grt
        #60                 disable wait_for_grt;
        @ (posedge grt) disable wait_for_grt;
    join
    good = (grt == 1'b1);
end
endtask
```

Consider all possible failure modes.

There is one problem with the models of the bus request operation in Sample 5-43 and Sample 5-44. What if the arbiter was functionally incorrect and left the *grt* signal always asserted? Both models would never see a rising edge on the *grt* signal. They would eventually exhaust their maximum waiting period then detect *grt* as asserted, indicating a successful completion. To detect this possible failure mode, the bus request operation must verify that the *grt* signal is not asserted prior to asserting the *req* signal, as shown in Sample 5-45.

Sample 5-45.
Verifying all failure modes in the asynchronous bus request operation

```
task bus_request;
   output good;
begin: bus_request_task
   if (grt == 1'b1) begin
      good = 1'b0;
      disable bus_request_task;
   end
   req = 1'b1;
   fork: wait_for_grt
      #60                     disable wait_for_grt;
      @ (posedge grt) disable wait_for_grt;
   join
   good = (grt == 1'b1);
end
endtask
```

Were you paying attention?

Pop quiz: The first *disable* statement in Sample 5-45 aborts the *bus_request* task and returns control to the calling block of the statement. Why does it disable the *begin/end* block inside the task and not the task itself?[5] And what is missing from all those task implementations?[6]

CPU Operations

Encapsulated operations are also known as bus-functional models.

Operations encapsulated using *tasks* or *procedures* can be very complex. The examples shown earlier were very simple, and dealt with only a few signals. Real life interfaces are more complex. But they can be encapsulated just as easily. These operations may even

---

5. For the answer see "Output Arguments on Disabled Tasks" on page 150.
6. They all include timing control statements. They should have a semaphore to detect concurrent activation. See "Non-Reentrant Tasks" on page 151.

return values to be verified against expected values or modify the stimulus sequence.

Test vectors "hard code" a number of wait states.

Figure 5-18 shows the specification for the read cycle for an Intel 386SX processor bus. Being a synchronous interface, it could be verified using test vectors. However, you would have to assume a specific number of wait cycles to sample the read data at the right time.

Figure 5-18.
Specification
for the read
cycle of a
386sx
processor

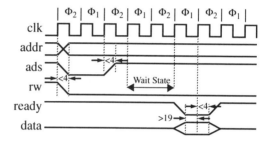

Bus models can adapt to a different number of wait states.

With behavioral models of the operation, you need not enforce a particular number of wait states and adapt to any valid bus timing. A model of the read operation can be found in Sample 5-46. The wait states are introduced by the fourth *wait* statement. How many failure modes are currently ignored by this model?[7]

Test vectors cannot perform read-modify-write operations.

In test vectors, the read value would have been specified as an expected output value. If that value had been different from the one specified, an error would have been detected. But what if you do not know the entire value that will be read? All you want is to modify the configuration of some slave device by reading its current configuration, modifying some bits, then writing the new configuration. This simple process is <u>impossible</u> to accomplish with test vectors blindly applied from a file to the inputs at every clock cycle. In behavioral testbenches, you can use the value returned during a read cycle, manipulate it, then use it in another operation. Sample 5-47 shows a portion of a testcase where the *read_cycle* procedure

7. Two: if clk = '1' and phi = '2' are never true and if *ready* is never asserted.

<table>
<tr>
<td>

Sample 5-46.
Model for the
read cycle
operation
</td>
<td>

```
procedure read_cycle (
          radd : in    std_logic_vector(0 to 23);
          rdat : out   std_logic_vector(0 to 31);
     signal clk  : in    std_logic;
     signal phi  : in    one_or_two;
     signal addr : out   std_logic_vector(0 to 23);
     signal ads  : out   std_logic;
     signal rw   : out   std_logic;
     signal ready: in    std_logic;
     signal data : inout std_logic_vector(0 to 31);
 is
 begin
     wait on clk until clk = '1' and phi = 2;
     addr <= radd after rnd_pkg.random * 4 ns;
     ads  <= '0'  after rnd_pkg.random * 4 ns;
     rw   <= '0'  after rnd_pkg.random * 4 ns;
     wait until clk = '1';
     wait until clk = '1';
     ads  <= '1'  after rnd_pkg.random * 4 ns;
     wait on clk until clk = '1' and phi = 2 and
                         ready = '0';
     assert ready'stable(19 ns) and
            data'stable(19 ns);
     rdat := data;
     wait for 4 ns;
     assert ready = '1' and data = (others => 'Z');
 end read_cycle;
```
</td>
</tr>
</table>

shown in Sample 5-46 and its corresponding *write_cycle* procedure
are used to perform a read-modify-write operation.

<table>
<tr>
<td>

Sample 5-47.
Performing a
read-modify-
write opera-
tion
</td>
<td>

```
test_procedure: process
    constant cfg_reg: std_logic_vector(0 to 23)
        := "000000000000001100010110";
    variable tmp: std_logic_vector(31 downto 0);
 begin
    ...
    i386sx_pkg.read_cycle(cfg_reg, tmp, ...);
    tmp(13 downto 9) := "01101";
    i386sx_pkg.write_cycle(cfg_reg, tmp, ...);
    ...
 end process test_procedure;
```
</td>
</tr>
</table>

Configurable Operations

Interfaces can have configurable elements.

An interface specification may contain configuration options. For example, the assertion level for a particular control signal may be configurable to either high or low. Each option has a small impact on the operation of the interface. Taken individually, you could create a different task or procedure for each configuration. The problem would be relegated to the testcase in deciding which flavor of the operation to invoke. You would also have to maintain several nearly-identical models.

Simple configurable elements become complex when grouped.

Taken together, the number of possible configurations explodes factorially.[8] It would be impractical to provide a different procedure or task for each possible configuration. It is much easier to include configurability in the model of the operation, and make the current configuration an additional parameter. An RS-232 interface, shown in Figure 5-19, is the perfect example of a highly configurable operation. Not only is the polarity of the parity bit configurable, but also its presence, as well as the number of data bits transmitted. And to top it all, because the interface is asynchronous, the duration of each pulse is also configurable. Assuming eight possible baud rates, five possible parities, seven or eight data bits, and one or two stop bits, there are 160 possible combinations of these four configurable parameters.

Figure 5-19. Specification for the RS-232 interface

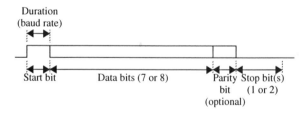

Duration (baud rate)

Start bit   Data bits (7 or 8)   Parity Stop bit(s)
                                     bit   (1 or 2)
                                  (optional)

Write a configurable operation model.

Instead of writing 160 flavors of the same operation, it is much easier to model the configurablity itself, as shown in Sample 5-48. The configuration parameter is assumed to be a record containing a field for each of the four parameters. Since Verilog does not directly support record types, refer to "Records" on page 105 for the implemen-

---

8. Exponential growth follows a K^n curve. Factorial growth follows a $n!$ curve, where $n! = 1 \times 2 \times 3 \times 4 \times ... \times (n-2) \times (n-1) \times n$.

tation details. What important safety measure is missing from the task in Sample 5-48?[9]

Sample 5-48.
Model for a
configurable
operation

```
'define sec * 1000000000  // timescale dependent!
task rs232_tx;
    input [7:0]              data;
    input 'rs232_cfg_typ cfg;

    time    duration;
    integer i;
begin
    duration = (1 'sec) / cfg'baud_rate;
    tx = 1'b1;
    #(duration);
    for (i = cfg'n_bits; i >= 0; i = i-1) begin
        tx = data[i];
        #(duration);
    end
    if (cfg'parity != 'none) begin
        if (cfg'n_bits == 7) data[7] = 1'b0;
        case (cfg'parity)
        'odd  : tx = ~^data;
        'even : tx = ^data;
        'mark : tx = 1'b1;
        'space: tx = 1'b0;
        endcase
        #(duration);
    end
    tx = 1'b0;
    repeat (cfg'n_stops) #(duration);
end
endtask
```

COMPLEX RESPONSE

Output verification
must be auto-
mated.

We have already established that visual inspection is not a viable option for verifying even a simple response. Complex responses are definitely not verifiable using visual inspection of waveforms. The process for verifying the output response must be automated.

---

9. The task contains timing control statements. It should contain a sema-phore to detect concurrent activation. See "Non-Reentrant Tasks" on page 151.

Verifying response is usually not as simple as checking the outputs after each stimulus operation. In this section, I describe how complex monitors are implemented using bus-functional models. I show how to manage separate control threads to make a testbench independent of the latency or delay within a design. I explain how to design a series of output monitors to handle non-deterministic responses as well as bi-directional interfaces.

What is a Complex Response?

Latency and output protocols create complex responses.

I define a complex response as something that cannot be verified in the same process that generates the stimulus. A complex response situation could be created simply because the validity of the output cannot be verified at a single point in time. It could also be created by a long (and potentially variable) delay between the stimulus and the corresponding response. These types of responses cannot be verified as part of the stimulus generation because the input sequence would be interrupted while it waits for the corresponding output value to appear. Interrupting the input sequence would prevent stressing the design at the fastest possible input rate. Holding the input sequence may even prevent the output from appearing or violate the input protocol. A complex response must be verified autonomously from the stimulus generation.

A simple design can have a complex response.

A Universal Asynchronous Receiver Transmitter (UART) is a perfect example of a simple design with a complex response. And not only because the output operation is configurable. Figure 5-20 shows the block diagram of the transmit path. Because the RS-232 protocol is so much slower than today's processor interfaces, waiting for the output corresponding to the last CPU write cycle would introduce huge gaps in the input stimulus, as shown in Figure 5-21. The design would definitely not be verified under maximum input stress conditions.

Figure 5-20.
Block diagram
of a UART
transmit path

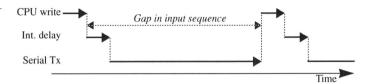

Figure 5-21.
Stimulus and
reponse
checking in a
single process

To stress the input of the design under maximum data rate condition, the testcase must decouple the input generation from the output verification. The stimulus generation part would issue write cycles as fast as it could, as long as the design can accept the data to be transmitted. It would stop generating write cycles only when the FIFO was full and the CPU interface signals that it can no longer accept data. The output would be verified in parallel, checking that the words that were sent to the design via the CPU interface were properly received via the serial interface. Figure 5-22 shows the timing of checking the output independently from generating the input. Gaps in the input sequence are created by the design's own inability to sustain the input rate, not by a limitation of the verification procedure.

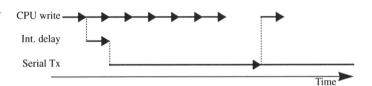

Figure 5-22.
Stimulus and
reponse
checking in
independent
processes

Handling Unknown or Variable Latency

Test vectors cannot deal with variable latency.

When using test vectors, you have to assume a specific delay between the input and its corresponding output. The delay is expressed in terms of the number of clock cycles it takes for the input value to be processed into its output value. This delay is known as the *latency* of the design. Latency is usually a by-product of the architecture of the design and is a side-effect of the pipelining required to meet the register-to-register timing constraints. The specific latency of a design is normally known only toward the very end of the RTL design process. A specific latency is rarely a design requirement. If a specific latency is not a requirement, why enforce one in the verification?

Verify only the characteristics that define functional correctness.	In the UART design from Figure 5-20, latency is introduced by the CPU interface, the FIFO and the serial interface. Externally, it translates into the internal delay shown in Figure 5-21 and Figure 5-22. The latency of the UART design is functionally irrelevant. The functional correctness of the design is solely determined by the data being transmitted, unmodified, in the same order in which it was received by the CPU interface. Those are the only criteria the testbench should be verifying. Any other limitations imposed by the testbench would either limit the freedom of choice for the RTL designer in implementing the design, or turn into a maintenance problem for you, the testbench designer.
Stimulus and response are implemented using different execution threads.	Verification of the output independently from the stimulus generation requires that each be implemented in separate *execution threads*. Each must execute independently from the other, i.e., in separate parallel constructs (*processes* in VHDL, *always* or *initial* blocks and *fork/join* statements in Verilog). These execution threads need to be synchronized at appropriate points. Synchronization is required to notify the response checking thread that the stimulus generation thread is entering a different phase of the test sequence. It is also required when either thread has completed its duty for this portion of the test sequence and the other thread can move on to the next phase.

| **Sample 5-49.** Using a named event in Verilog | ```
event sync;
initial
begin: stimulus
    . . .
    -> sync;
    . . .
end

initial
begin: response
    . . .
    @ (sync);
    . . .
end
``` |
| Synchronize threads using *fork/join* or named events in Verilog or signal activity in VHDL. | In Verilog, implicit synchronization occurs when using the *fork/join* statement. Explicit synchronization is implemented using the *named event*, as illustrated in Sample 5-49. In VHDL, synchronization is implemented using a toggling signal, as shown in Sample 5-50. The actual value of the boolean signal is irrelevant. The infor- |
```

mation is in the timing of the value-change. Alternatively, the *'transaction* signal attribute can be used to synchronize a process with the assignment of a value to a signal, as shown in Sample 5-51. Pop quiz: why use the *'transaction* attribute and not simply wait for the event on the signal caused by the new value?[10]

Sample 5-50.
Using a toggling boolean in VHDL

```
architecture test of bench is
   signal sync: boolean;
begin
   stimulus: process
   begin
      . . .
      sync <= not sync;
      . . .
   end process stimulus;

   response: process
   begin
      . . .
      wait on sync;
      . . .
   end process response;
end test;
```

Sample 5-51.
Using the *'transaction* attribute in VHDL

```
architecture test of bench is
   signal expect: integer;
begin
   stimulus: process
   begin
      . . .
      expect <= ...;
      . . .
   end process stimulus;

   response: process
   begin
      . . .
      wait on expect'transaction;
      . . .
   end process response;
end test;
```

10. Because the assignment of the same value, twice in a row, will not cause an event on the signal. To synchronize to the assignment of <u>any</u> value on a signal, you must be sensitive to any assignment, even of values that do not cause an event. That's what the *'transaction* attribute identifies.

Figure 5-23 shows the execution threads for the verification of the UART transmit path. It also shows the synchronization points when the testcase switches from loading a new configuration in the UART to actively transmitting data, and vice-versa.

Figure 5-23.
Execution
threads for
UART Tx
testcase

The Verilog implementation of the execution threads shown in Figure 5-23 is detailed in Sample 5-52.

Sample 5-52.
Implementing
execution
threads in Ver-
ilog

```
initial
begin
    ... // Init simulation

    fork: config_phase
        begin
            ... // Config
            disable config_phase
        end
        begin
            ... // Check output remains idle
        end
    join

    fork: data_phase
        begin
            ... // Write data to send via CPU i/f
        end
        begin
            ... // Check data sent serially
        end
    join

    ... // Terminate simulation
end
```

Controlling execu-
tion threads is sim-
plified by using
the *fork/join* state-
ment.

The implementation in VHDL is a little more complex. Since VHDL lacks the *fork/join* statement, individual processes must be used. A process must be selected as the "master" process. The master process controls the synchronization of the various execution threads in the testcase. The master process can be a separate process whose sole function is to control the execution of the testcase. It could also be one of the execution threads, usually one of the stimulus generation threads. An emulation of the *fork/join* statement, as shown in "Fork/Join Statement" on page 134, could be used. In Sample 5-53, a simple synchronization scheme using a different signal for each synchronization point is used.

Sample 5-53.
Implementing
execution
threads in
VHDL

```
architecture test of bench is
    signal sync1, syn2, sync3, done: boolean;
begin
    stimulus: process
    begin
        ... -- Init simulation
        sync1 <= not sync1;
        ... loop
            ... -- Config via CPU i/f
            sync2 <= not sync2;
            ... -- Write data to send via CPU i/f;
            -- Wait for data to be received
            wait on sync3;
        end loop;
        done <= true;
        wait;
    end process stimulus;

    response: process
    begin
        -- Wait until init is complete
        wait on sync1;
        loop
            -- Check output is idle while config
            wait until Tx /= '0' or
                        sync2'event or done;
            if done then
                -- Terminate simulation
                assert FALSE severity FAILURE;
            end if;
            assert Tx = '0' ...;
            ... -- Verify data sent serially
            sync3 <= not sync3;
        end loop;
    end process response;
end test;
```

Abstracting Output Operations

Output operations can be encapsulated.

Earlier in this chapter, we encapsulated input operations to abstract the stimulus generation from individual signals and waveforms to generating sequences of operations. A similar abstraction can be used for verifying the output. The repetitiveness of output signals is taken care of and verified inside the subprograms. The output verification thread simply passes the expected output value to the monitor subprogram.

Arguments include expected value and configuration parameters.

The input operations takes as argument the specific value to use to generate the stimulus. Conversely, the output operations, encapsulated using *tasks* in Verilog, or *procedures* in VHDL, take as argumen the value expected to be produced by the design. If the format or protocol of the output operation is configurable, its implementation should be configurable as well.

A perfect example is the operation to verify the serial output in a UART transmit path, shown in Figure 5-20. It is as highly configurable as its input counterpart, detailed in Sample 5-48. The difference is that it compares the specified value with the one received, and compares the received parity against its expected value based on the received data and the configuration parameters. An implementation of the RS-232 receiver operation is detailed in Sample 5-54. It assumes that the configuration is specified using a user-

Sample 5-54. Implementation of the RS-232 serial receive operation

```
subtype byte is std_logic_vector(7 downto 0);
procedure recv(signal rx    : in std_logic;
                      expect: in byte;
                      config: in rs232_cfg_typ)
   is
      variable period: time;
      variable actual: byte := (others => '0');
   begin
      period := 1 sec / config.baud_rate;
      wait until rx = '1';       -- Wait for start bit
      wait for period / 2;       -- Sample mid-pulse
      for I in config.n_bits downto 0 loop
         wait for period;
         actual(I) := rx;        -- 7-8 data bits
      end loop;
      assert actual = expect; -- Compare
      -- Parity bit?
      if (config.parity /= no_parity) then
         wait for period;
         assert rx = parity(actual, config.parity);
      end
      wait for period;           -- Stop bit
      assert rx = '0';
   end recv;
```

defined record type. A function to compute the parity of an array of bits, based on a configurable parity value, is also assumed to exist. This parity function could use the *xor_reduce* function available in Synopsys's *std_logic_misc* package.

Consider all possible failure modes.

The procedure shown in Sample 5-54 has some potential problems and limitations. What if the output being monitored is dead and the start bit is never received? This procedure will hang forever. It may be a good idea to provide a maximum delay to wait for the start bit via an additional argument, as shown in Sample 5-55, or to compute a sensible maximum delay based on the baud rate. Notice how a default argument value is used in the procedure definition to avoid forcing the user to specify a value when it is not relevant, as shown in Sample 5-56, or to avoid modifying existing code that was written before the additional argument was added.

Sample 5-55. Providing an optional timeout for the RS-232 serial receive operation

```
procedure recv(signal rx     : in std_logic;
                      expect : in byte;
                      config : in rs232_cfg_typ:
                      timeout: in time
                                    := TIME'high)
is
   ...
begin
   ..
   wait until rx = '1' for timeout;
   assert rx = '1';
   ...
end recv;
```

Sample 5-56. Using the RS-232 serial receive operation

```
process
begin
   ...
   recv(rx, "01010101", cfg_9600_8N1, 100 ms);
   recv(rx, "10101010", cfg_9600_8N1);
   ...
end process;
```

Do not arbitrarily constrain the operation.

The width of pulses is not verified in the implementation of the RS-232 receive operation in Sample 5-54. Should it? If you assume that the procedure is used in a controlled, 100 percent digital environment, then verifying the pulse width might make sense. This procedure could also be used in system-level verification, where the serial signal was digitized from a noisy analog transmission line as illustrated in Figure 5-24. In that environment, the shape of the pulse, although unambiguously carrying valid data, most likely does not meet the rigid requirements of a clean waveform for a specific baud rate. Just as in real life, where modems fail to communicate properly if their baud rates are not compatible, an improper waveform shape is detected as invalid data being transmitted.

Figure 5-24.
Modification
to the serial
signal in a real
system

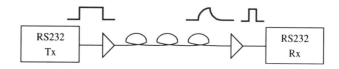

Generic Output Monitors

Verifying the value in the output monitor is too restrictive.

The output verification operation, as encapsulated in Sample 5-54, has a very limited application. It can be only used to verify that the output value matches a predefined expected value. Can you imag-

Sample 5-57.
Generic RS-
232 serial
receive opera-
tion

```
subtype byte is std_logic_vector(7 downto 0);
procedure recv(signal rx    : in   std_logic;
                      actual: out byte;
                      config: in   rs232_cfg_typ)
is
    variable period: time;
    variable data  : byte;
begin
    period := 1 sec / config.baud_rate;
    wait until rx = '1';      -- Wait for start bit
    wait for period / 2;      -- Sample mid-pulse
    data(7) := '0';           -- Handle 7 data bits
    for I in config.n_bits downto 0 loop
        wait for period;
        data(I) := rx;        -- 7-8 data bits
    end loop;
    -- Parity bit?
    if (config.parity /= no_parity) then
        wait for period;
        assert rx = parity(data, config.parity);
    end
    wait for period;          -- Stop bit
    assert rx = '0';
    actual := data;
end recv;
```

ine other possible uses? What if the output value can be any value within a predetermined set or range? What if the output value is to be ignored until a specific sequence of output values is seen? What if the output value, once verified, needs to be fed back to the stimulus generation? The usage possibilities are endless. It is not possible, a priori, to determine all of them nor to provide a single interface that satisfies all of their needs.

Separate monitoring from value verification.

The most flexible implementation for an output operation monitor is to simply return to the caller whatever output value was just received. It will be up to a "higher authority" to determine if this value is correct or not. The RS-232 receiver was modified in Sample 5-57 to return the byte received without verifying its correctness. The parity, being independent of the correctness of the value and fully contained within the RS-232 procotol, can still be verified in the procedure.

Monitoring Multiple Possible Operations

The next operation on an output interface may not be predictable.

You may be in a situation where more than one type of operation can happen on an output interface. Each would be valid and you cannot predict which specific operation will come next. An example would be a processor that executes instructions out of order. You cannot predict (without detailed knowledge of the processor architecture) whether a read or a write cycle will appear next on the data memory interface. The functional validity is determined by the proper access sequence to related data locations.

Sample 5-58.
Processor test program

```
load A,  R0
load B,  R1
add   R0, R1, R2
sto   R2, X
load C,  R3
add   R0, R3, R4
sto   R4, Y
```

Verify the sequence of related operations.

For example, consider the testcase composed of the instructions in Sample 5-58. It has many possible execution orders. From the perspective of the data memory, the execution is valid if the conditions listed below are true.

1. Location A is read before location X and Y are written.

2. Location B is read before location X is written.

3. Location C is read before location Y is written.

4. Location X must be written with the value A+B.

5. Location Y must be written with the value A+C.

These are the sufficient and necessary conditions for a proper execution of the test program. Verifying for a particular order of the individual cycle overconstrains the testcase.

Write an operation "dispatcher" task or procedure.

How do you write an encapsulated output monitor when you do not know what kind of operation comes next? You must first write a monitor that identifies the next cycle after it has started. It verifies the preamble to all operations on the output interface until it becomes unique to a specific operation. It then returns any information collected so far and identifies, to the testbench, which cycle is currently underway. It is up to the testbench to then call the appropriate *task* or *procedure* to complete the verification of the operation.

Sample 5-59 shows the skeleton of a monitor task that identifies whether the next operation for a CPU is a read or a write cycle. Since the address has already been sampled by the time the decision of the type of cycle was made, it is returned along with the current cycle type. Sample 5-60 shows how this operation identification task is used by the testbench to determine the next course of action.

Sample 5-59.
Monitoring many possible output operations

```
parameter READ_CYCLE  = 0,
          WRITE_CYCLE = 1;
time last_addr;
task next_cycle_is;
    output          cycle_kind;
    output [23:0] address;
begin
    @ (negedge ale);
    address = addr;
    cycle_kind =
        (rw == 1'b1) ? READ_CYCLE : WRITE_CYCLE;
    #(Tahold);
    if ($time - last_addr < Tahold + Tasetup)
        $write("Setup/Hold time viol. on addr\n");
end
endtask

always @ (addr) last_addr = $time;
```

<table>
<tr>
<td>

Sample 5-60.
Handling
many possible
output opera-
tions

</td>
<td>

```
initial
begin: test_procedure
    reg          cycle_kind;
    reg [23:0] addr;

    next_cycle_is(cycle_kind, addr);
    case (cycle_kind)
    READ_CYCLE:  read_cycle(addr);
    WRITE_CYCLE: write_cycle(addr);
    endcase
    ...
end
```

</td>
</tr>
</table>

In this case, we assume the existence of two tasks, one for each possible operation, which completes the monitoring of the remainder of the cycle currently under way.

Monitoring Bi-Directional Interfaces

Output interfaces may need to reply with "input" data.

We have already seen that input operations sometimes have to monitor some signals from the design under verification. The same is true for output monitor. Sometimes, they have to provide data back as an answer to an "output" operation. This blurs the line between stimulus and response. Isn't a stimulus generation subprogram that verifies the control or feedback signals from the design also doing response checking? Isn't a monitor subprogram that replies with new data back to the design also doing stimulus generation?

Generation and monitoring pertains to the ability to initiate an operation.

The terms *generator* and *monitor* become meaningless if they are attached to the direction of the signals being generated or monitored. They regain their meaning if you attach them to the initiation of operations. If a procedure or a task *initiates* the operation, it is a *stimulus generator*. If the procedure or task sits there and waits for an operation *to be initiated by the design*, then it is an *output monitor*. The latter also includes ancilliary tasks to complete a cycle currently underway, as discussed in "Monitoring Multiple Possible Operations" on page 203.

Bridges have bi-directional output interfaces.

The downstream master interface on a bus bridge is the perfect example of a bi-directional "output" interface. The bridge is the design under verification. An example, illustrated in Figure 5-25, is a bridge between a proprietary on-chip bus and a PCI interface. The cycles are initiated on the on-chip bus (upstream). If the address falls within the bridge's address space, it translates the cycle onto the PCI bus (downstream). This bridge allows master devices on the on-chip bus to transparently access slave devices on the PCI bus.

Figure 5-25.
Bridge between an on-chip bus and a PCI bus

Using a memory to test a CPU interface can mask certain classes of problems.

To verify this bridge, you would need an on-chip bus cycle generator and a PCI bus cycle monitor, as illustrated in Figure 5-26. Many would be tempted to use a model of a memory (which, for PCI, are readily available from model suppliers), instead of a PCI monitor. The verification would be accomplished by writing a pattern in the memory then reading it back. Using a memory would not catch several types of problems masked by the readback operations.

For example, what if the bridge designer misreads the PCI specification document and implements the address bus using little endian instead of big endian? During the write cycle, address 0xDEADBEEF on the on-chip bus is translated to the physical address 0xEFBEADDE on the PCI bus, writing the data in the wrong memory location. The read cycle, used to verify that the write cycle is correct, also translates address 0XDEADBEEF to 0xEFBEADDE and reads the data from the <u>same</u>, but <u>invalid</u> location. The testbench does not have the necessary visibility to detect the error.

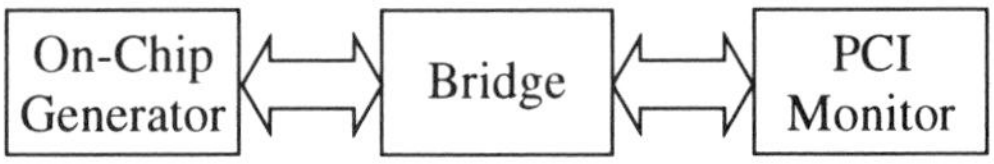

Figure 5-26.
Verification structure for the bridge

On-Chip Generator — Bridge — PCI Monitor

Use a monitor that detects the PCI cycles and notifies the testbench.

Using a generic PCI monitor to verify the output detects errors that would be masked by using a write-readback process. The PCI monitor task or procedure would watch the bus until it determined the type of cycle being initiated by the bridge. To ease implementation, this task or procedure usually continues to monitor the bus while the cycles remain identical (i.e. for the entire address phase). Assuming that the bridge's implementation is limited to generating

Assuming that the bridge's implementation is limited to generating PCI memory read and write cycles, the monitor task or procedure would then return, identifying the cycle as a memory read or write and the address being read or written. The skeleton for a PCI bus monitor is shown in Sample 5-61.

Sample 5-61.
Monitoring
many possible
PCI cycles

```
parameter MEM_RD = 0,
          MEM_WR = 1;
task next_pci_cycle_is;
    output          cycle_kind;
    ouptut [31:0] address;
begin
    // Wait for the start of the cycle
    @ (posedge pci_clk);
    while (frame_n !== 1'b0) @ (posedge pci_clk);
    // Sample command and address
    case (cbe_n)
    4'b0110: cycle_kind = MEM_RD;
    4'b0111: cycle_kind = MEM_WR;
    default: $write("Unexpected cycle type!\n");
    endcase
    address = ad;
end
endtask
```

You must be able to verify different possible answers by the bus monitor.

The really interesting part comes next. In PCI, read and write cycles can handle arbitrary length bursts of multiple data values in a single cycle. A read cycle can read any number of consecutive bytes and a write cycle can write any number of consecutive bytes. The PCI master is under control of the length of the burst, but the number of bytes involved in a cycle is <u>not</u> specified in the preamble. Data must be read or written by the slave for as long as the master keeps reading or writing them.

In a VHDL procedure, implementing a monitor for the write cycle, you could use an access value to an array of bytes to return all of the data values that were written during a burst. The instance of the array object would be dynamically allocated with the proper constraints according to the number of bytes read. An example is shown in Sample 5-62. But what about read cycles where the testbench cannot know, a priori, how many bytes will be read? And what about Verilog which does not support arrays of bytes on interfaces, let alone unconstrained arrays?

Sample 5-62.
Monitoring
burst write
cycles in
VHDL

```
subtype byte is std_logic_vector(7 downto 0);
type byte_array_typ is array(natural range <>)
   of byte;
type burst_data_typ is access byte_array_typ;

procedure next_pci_cycle_is(...);

procedure pci_mem_write_cycle(
   -- PCI bus interface as signal class formals
   signal  pci_clk: in  std_locic;
   ...
   -- Pointer to data values written during cycle
   variable data_wr: out burst_data_typ);
```

A monitor can be composed of several *tasks* or *procedures* that must be appriopriately called by the testbench.

The PCI bus monitor has been implemented by slicing the PCI cycles into at least two procedures: one to handle the generic preamble and detect the type of cycle under way, the other intended to handle the remainder of each cycle.

The solution to our dilemma is to slice the implementation of the PCI bus monitor even further: use a procedure to handle <u>each data transfer</u> and one to handle the cycle termination.

The data transfer procedure or task would have an input or output argument for the data read or written, and an output argument to indicate whether to continue with more data transfers or terminate the cycle. Figure 5-27 illustrates how each procedure is sequenced, under the control of the testbench, to form a complete PCI cycle. A similar slicing strategy would be used in creating a PCI bus generator. The only exception being that the generator is now under the control of the initiation of the cycle and its duration.

Figure 5-27.
Slicing the
PCI cycle into
procedures

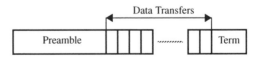

Provide the controls at the proper level of granularity.

Slicing the implementation of the PCI cycle at the data transfer level offers an additional opportunity. In PCI, both master and slave can throttle the transfer rate by asserting the *irdy_n* and *trdy_n* signals, respectively, when they are ready to complete a transfer.

A procedure or a task implementing a single data transfer can have an additional parameter specifying the number of clock cycles to wait before asserting the *trdy_n* signal. Another could be used to specify whether to force one of the target termination exception. It can also report that a master-initiated exception occured during this data transfer, such as a timeout, a system error, or a parity error.

The testbench would then be free to "generate" any number of possible answers to a PCI cycle. With the tasks outlined in Sample 5-61 and Sample 5-63, the possibilities become endless! One of these possibilities is shown in Sample 5-64.

Sample 5-63.
PCI data transfer and termination tasks

```
// Target terminations
parameter NORMAL     = 0,
          RETRY      = 1,
          DISCONNECT = 2,
          ABORT      = 3;
// Output status
parameter TERMINATE       = 0,
          CONTINUE        = 1,
          INITIATOR_ABORT = 2,
          PARITY_ERROR    = 3,
          SYS_ERROR       = 4;
task pci_data_rd_xfer;
   input  [31:0] rd_data;
   input  [ 7:0] delay;
   input  [ 1:0] termination;
   output [ 2:0] status;
   ...
endtask

task pci_data_wr_xfer;
   output [31:0] wr_data;
   input  [ 7:0] delay;
   input  [ 1:0] termination;
   output [ 2:0] status;
   ...
endtask

task pci_end_cycle;
   output [ 2:0] status;
   ...
endtask
```

```
initial
begin: test_procedure
    fork
        begin: on_chip_side
            // Generate a long read cycle on the
            // On-Chip bus side
            ...
        end
        begin: pci_side
            reg       kind;
            reg [31:0] addr;
            integer   delay;
            integer   ok;

            // Expect a read cycle on the PCI side
            // at the proper address
            next_pci_cycle_is(kind, addr);
            if (kind != MEM_RD) ...
            if (addr != ...) ...

            // Send back 5 random data words
            // with increasing delays in readiness
            // then force a target abort on the 6th.
            delay = 0;
            repeat (5) begin
                pci_data_rd_xfer($random, delay,
                                 NORMAL, ok);
                if (ok !== CONTINUE) ...
                delay = delay + 1;
            end
            pci_data_rd_xfer($random, 0, ABORT, ok);
        end
    join
end
```

Using a monitor
simplifies the
testcase.

Using the generic PCI bus monitor also shortens the testcase compared to using a memory. With the monitor, you have direct access
to all of the bus values. It is not necessary to write into the memory
for the entire range of address and data values, creating interesting
test patterns that highlight potential errors. With a monitor, only a
few addresses and data values are sufficient to verify that the bridge
properly translates them. It is also extremely difficult to control the
answers provided by the memory to test how the bridge reacts to
bus exceptions. These exception conditions become easy to setup
with a generic monitor designed to create them.

PREDICTING THE OUTPUT

The unstated assumption in implementing self-checking test-benches is that you have detailed knowledge of the output to be expected. Knowing exactly which output to expect and how it can be verified to determine functional correctness is the most crucial step in verification. In some cases, such as RAMs or ROMs, the response is easy to determine. In others, such as a video compressor or a speech synthesizer, the response is much more difficult to define. This section examines various families of designs and show how the expected response could be determined and communicated to the output monitors.

Data Formatters

The expected output equals the input.

There is a class of designs where the input information is not transformed, but simply reformated. Examples include UARTs, bridges, and FIFOs. They have the simplest output prediction process. Since the information is not modified, predicting the output is a simple matter of knowing the sequence of input values.

Forwarding one value at a time under-constrains the design.

Passing data values, one at a time, from the stimulus generator to the response monitor, as illustrated in Figure 5-28, is usually not appropriate. This limits the data rate to one every input and output cycle and may not stress the design under worse conditions. Pipelined designs cannot be verified using this stategy: input must be continuously supplied while their corresponding response has not yet appeared on the output.

Figure 5-28.
Forwarding
one value at a
time

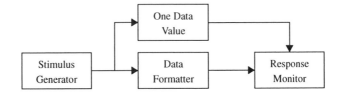

A short data sequence can be implemented in a global array.

If the input sequence is short and predetermined, using a global data sequence table is the simplest approach. Both the input generator and output monitor use the global table. The input generator applies each value in sequence. The output monitor compares each output value against the sequence of values in the global table. Figure 5-29 illustrates the flow of information while Sample 5-65 shows an implementation in VHDL.

Figure 5-29.
Using a global table to predict output.

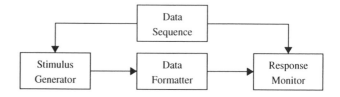

Sample 5-65.
Implementation of a global data sequence table

```
architecture test of bench is
    type std_lv_ary_typ is array(natural range <>)
        of std_logic_vector(7 downto 0);
    constant walking_ones: std_lv_ary_typ(1 to 8)
        := ("10000000",
            "01000000",
            "00100000",
            "00010000",
            "00001000",
            "00000100",
            "00000010",
            "00000001");
begin
    DUV: ...

    stimulus: process
    begin
        for I in walking_ones'range loop
            apply(walking_ones(I), ...);
        end loop;
        wait;
    end process stimulus;

    response: process
    begin
        for I in walking_ones'range loop
            expect(walking_ones(I), ...);
        end loop;
        assert false severity failure;
    end process response;
end test;
```

Long data
sequence can use a
FIFO between the
generator and
monitor.

Often the input sequence is long or computed on-the-fly. It is not practical for hardcoding in a global constant. A FIFO can be used to forward expected values from the stimulus generator to the output monitor. The input generator puts each value in sequence in one end of the FIFO. The output monitor compares each output value against the sequence of values dequeued from the other end of the FIFO. This strategy is a simple extension of the concept of forward-

Figure 5-30.
Forwarding
data values via
a FIFO

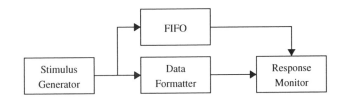

Sample 5-66.
Implementa-
tion using a
FIFO to for-
ward data val-
ues

```
task put_fifo;
   ...
endtask

function [7:0] get_fifo;
   ...
endfunction

initial
begin: stimulus
   reg [7:0] data;

   repeat (...) begin
      data = ...;
      put_fifo(data);
      apply(data);
   end
end

initial
begin: response
   reg [7:0] data;

   repeat (...) begin
      data = get_fifo(0);
      expect(data);
   end
   $finish;
end
```

ing a single value at a time. It is illustrated in Figure 5-30. Notice how the architecture of the testbench is identical to the one illustrated in Figure 5-28. The code in Sample 5-66 shows the implementation structure in Verilog of a testbench using a FIFO. The implementation of the FIFO itself is left as an exercise to the reader.

The stimulus and response processes can read the same data file.

Sometimes, the data sequence is externally generated and supplied to the testbench using a data file. A file can be read, concurrently, by more than one process. Thus, the stimulus generator and response monitor can both read the file, using it in a fashion similar to a global array. The code in Sample 5-67 illustrates how this strategy could be implemented in VHDL. The filename is assumed to be

Sample 5-67.
Implementation using an external data file

```
entity bench is
    generic (datafile: string);
end bench;

architecture test of bench is
begin
    DUV: ...

    stimulus: process
        file infile : text is in datafile;
        variable L   : line;
        variable dat: std_logic_vector(7 downto 0);
    begin
        while not endfile(infile) loop
            readline(infile, L);
            read(L, dat);
            apply(dat, ...);
        end loop;
        wait;
    end process stimulus;

    response: process
        file infile : text is in datafile;
        variable L   : line;
        variable dat: std_logic_vector(7 downto 0);
    begin
        while not endfile(infile) loop
            readline(infile, L);
            read(L, dat);
            expect(dat, ...);
        end loop;
        assert false severity failure;
    end process response;
end test;
```

passed to the testbench via the command line using a generic of type *string*.

Packet Processors

Packets have untouched data fields.

This family of designs uses some of the input information for processing, sometimes transforming it. But it leaves portions of the input untouched and forwards it, intact, all the way through the design to an output. Examples abound in the datacom industry. They include Ethernet hubs, IP routers, ATM switches, and SONET framers.

Use the untouched fields to encode the expected tranformation.

The portion of the data input that passes, untouched, through the design under verification can be put to good use. It is often called *payload* and the term *packet* or *frame* is often used to describe the unit of data processed by the design. You must first determine, through a proper testcase, that the payload information is indeed not modified by the design. Subsequently, it can be used to describe the expected output for this packet. For each packet received, the output monitor uses the information in the payload to determine if it was appropriately processed.

This simplifies the testbench control structure.

Figure 5-31 shows the structure of a testbench for a four-input and four-output packet router. Notice how the output monitors are completely autonomous. This type of design usually lends itself to the simplest testbench control structures, assuming that the output monitors are sufficiently intelligent. The control of this type of testbench is simple because all the processing (stimulus and generation of expected response) is performed in a single location: the stimulus generator. Some minor orchestration between the generators may be required in some testcases when it is necessary to synchronize traffic patterns to create interesting scenarios.

Figure 5-31.
Testbench structure for a 4x4 packet router

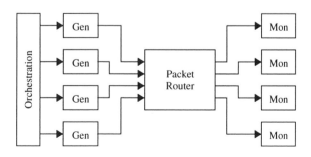

Include all necessary information in the payload to determine functional correctness.

The payload must contain all necessary information to determine if a particular packet came out of the appropriate output, and with the appropriate transformation of its control information.

For example, assume the success criteria is that the packets for a given input stream be received in the proper order by the proper output port. The payload should contain a unique stream identifier, a sequence number, and an output port identifier, as shown in Figure 5-32.

The output monitor needs to verify that the output identifier matches its own identifer. It also needs to verify that the sequence number is equal to the previously-received sequence number in that stream plus one, as outlined in Sample 5-68. The Verilog records are assumed to be implemented using the technique shown in "Records" on page 105.

Figure 5-32. Example packet payload structure

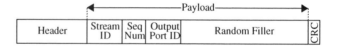

Sample 5-68. Implementation using payload information to determine functional correctness

```
always
begin: monitor
    reg 'packet_typ pkt;

    receive_packet(pkt);
    // Packet is for this port?
    if (pkt'out_port_id !== my_id) ...;
    // Packet in correct sequence?
    if (last_seq[pkt'strm_id] + 1
        != pkt'seq_num) ...;
    // Reset sequence number
    last_seq[pkt'strm_id] = pkt'seq_num;
end
```

Complex Transformations

The last family of designs processes and transforms the input data completely and thoroughly. The expected output can be only determined by reproducing the transformation using alternative means. This includes reversing the process where you determine which input sequence to provide in order to produce a desired output.

Use a truly alternative computation method.

When reproducing the transformation, to determine which output value to expect, as illustrated in Figure 5-33, you must use a different implementation of the transformation algorithm. For example, you can use a reference C model. For a DSP implementation, you could use floating-point expressions and the predefined *real* data types to duplicate the processing that is performed using fixed-point operators and data representation in the design.

Figure 5-33. Reproducing the transformation to predict output

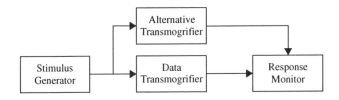

Use a different programming model for the output monitor.

If you are providing an input sequence to produce a specific output pattern, use a different programming model for the output monitor. The programming model for the design was chosen to ease implementation - or even to make it possible. Having almost no constraints in behavioral HDL models, you can choose a programming model that is more natural, to express the expected output. Using a different programming model also forces your mind to work in a different way when specifying the input and output, creating an alternative verification path.

Figure 5-34. Target waveform to generate

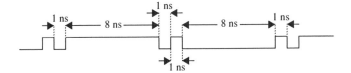

Example: a waveform generator.

For example, you could be verifying a design that generates arbitrary digital waveforms. The input could specify, for each clock cycle, the position of up to three rising or falling edges within the clock period. Each transition is specified using two parameters. A *level* bit indicates the final logic level after the transition and a 10-bit *offset* value indicates the position of the transitions within the 10 ns clock period, with a resolution of 9.7 ps (or 10 ns / 1024). Assuming that the waveform in Figure 5-34 represents an interesting testcase, Figure 5-35 shows how it is sliced to create the input sequence and Sample 5-69 shows how the stimulus could be generated.

Figure 5-35.
Slices for
input
specification

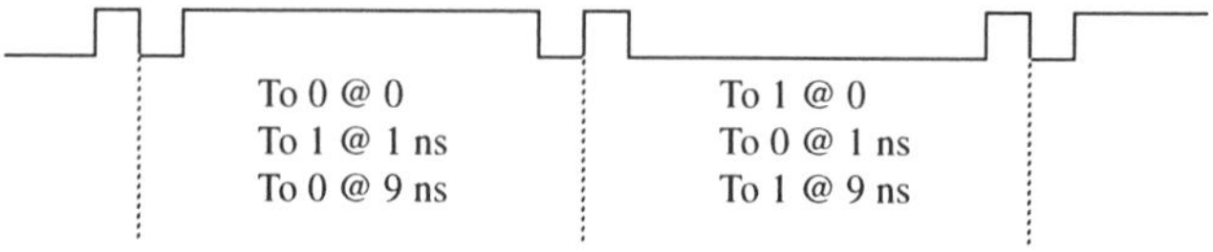

To 0 @ 0	To 1 @ 0
To 1 @ 1 ns	To 0 @ 1 ns
To 0 @ 9 ns	To 1 @ 9 ns

Sample 5-69.
Generating the
input for the
waveform
generator

```
initial
begin: stimulus
   repeat (10) begin
      apply(1'b0, $realtobits(0.0),
            1'b1, $realtobits(1.0),
            1'b0, $realtobits(9.0));
      apply(1'b1, $realtobits(0.0),
            1'b0, $realtobits(1.0),
            1'b1, $realtobits(9.0));
   end
end
```

Choose a different
but reliable way of
representing the
output.

How should the output be represented? If we use a slicing method similar to the input's, it would not provide for an alternative programming model. Furthermore, the implementation could miss transitions or events between slices. The output waveform has no relationship with the clock. Trying to specify the expected output using clock-based slices would simply over-constrain the test. The validity of the output waveform is entirely contained in the relative position of the edges. So why not specify the expected output using

Sample 5-70.
Monitoring the
generated
waveform

```
monitor: process
begin
   wait until wave = '1';
   for I in 1 to 10 loop
      wait on wave;
      assert wave'delayed'last_event = 8 ns;
      wait on wave;
      assert wave'delayed'last_event = 1 ns;
      wait on wave;
      assert wave'delayed'last_event = 1 ns;
   end loop;
   assert false severity failure;
end process monitor;
```

an edge-to-edge specification? Assuming that the output is initialized to a level of '0', an implementation of the output monitor is shown in Sample 5-70.

The 'delayed attribute must be used to look before the wait statement.

The 'delayed signal attribute must be used, otherwise 'last_event always returns 0 since the signal *wave* just had an event to resume the execution of the previous *wait* statement. The 'delayed attribute delays the *wave* signal by one delta cycle. The delayed *wave* signals looks to the 'last_event attribute as if it were *before* the execution of the *wait* statement. Notice how this monitor simply waits for the first rising edge of the output monitor to anchor its edge-to-edge relationships. This makes the monitor completely independent of the latency and intrinsic delays in the design.

Do not enforce unnecessary precision.

There is one problem with the verification of the delay between edges in Sample 5-70. Each delay is compared to a precise value. However, the design has a resolution of 9.7 ps. Each delay is valid if it falls in the range of the ideal delay value plus or minus the resolution, as shown in Sample 5-71.

Sample 5-71.
Handling
uncertainty in
the generated
waveform

```
monitor: process
    function near(val, ref: in time)
        return boolean is
        constant resolution: time := 9700 fs;
    begin
        return ref - resolution <= val and
                val <= ref + resolution;
    end near;
begin
    wait until wave = '1';
    for I in 1 to 10 loop
        wait on wave;
        assert near(wave'delayed'last_event, 8 ns);
        wait on wave;
        assert near(wave'delayed'last_event, 1 ns);
        wait on wave;
        assert near(wave'delayed'last_event, 1 ns);
    end loop;
    assert false severity failure;
end process monitor;
```

SUMMARY

In this chapter, I have described how to use bus-functional models to generate stimulus and monitor response. The bus-functional models were used to translate between high-level data representations and physical implementation levels. They also abstracted the interface operations, removing the testcases from the detailed implementation of each physical interface. Some of these bus-functional models can be very complex, depending on feed-back from the device under verification to operate properly or having to supply handshake information back to the device.

This chapter, after highlighting the problems with visual inpection, also described how to make each individual testbench completely self-checking. The expected response must be embedded in the testbench at the same time as the stimulus. Various strategies for determining the expected response and communicating it to the output monitors have been presented.

CHAPTER 6   ARCHITECTING TESTBENCHES

A testbench need not be a monolithic block. Although Figure 1-1 shows the testbench as a big thing that surrounds the design under verification, it need not be implemented that way. The design is also shown in a single block and it is surely not implemented as a single unit. Why should the testbench by any different?

The previous chapter was about low-level testbench components.

In Chapter 5, we focused on the generation and monitoring of the low-level signals going into and coming out of the device under verification. I showed how to abstract them into operations using bus-functional models. Each were implemented using a *procedure* or a *task*. The emphasis was on the stimulus and response of interfaces and the need for managing separate execution threads. If you prefer a bottom-up approach to writing testbenches, I suggest you start with the previous chapter.

This chapter focuses on the structure of the testbench.

This chapter concentrates on implementing the many testbenches that were identified in your verification plan. I show how to best structure the stimulus generators and response monitors to minimize maintenance, facilitate implementing a large number of testbenches, and promote the reusability of verification components.

REUSABLE VERIFICATION COMPONENTS

This section describes how to plan the architecture of testbenches. The goal is to maximize the amount of verification code reused across testbenches to minimize the development effort. The test-

Sample 6-1.
Implementing
the muxed
flip-flop test-
bench in Ver-
ilog

```verilog
module testbench;

reg rst, d0, d1, sel, clk;
wire q, qb;

muxed_ff duv(d0, d1, sel, q, qb, clk, rst);

parameter cycle  = 100,
          Tsetup =  15,
          Thold  =   5;

always
begin
   #(cycle/2) clk = 1'b0;
   #(cycle/2) clk = 1'b1;
end

task sync_reset;
   ...
endtask

task load_d0;
   input data;
begin
   rst <= 1'b0;
   d0  <= data;
   d1  <= ~data;
   sel <= 1'b0;
   @ (posedge clk);
   #(Thold);
   if (q !== data || qb !== ~data) ...
   {rst, d0, d1, sel} <= 4'bxxxx;
   #(cycle - Thold - Tsetup);
end
endtask

task load_d1;
   ...
endtask

initial
begin: test_sequence
   sync_reset;
   load_d0(1'b1);
   ...
   $finish;
end
endmodule
```

Sample 6-2.
Implementing
the muxed
flip-flop test-
bench in
VHDL

```
architecture test of bench is
    signal rst, d0, d1, sel, q, qb: std_logic;
    signal clk: std_logic := '0';
    component muxed_ff
        ...
    end component;
    constant cycle : time := 100 ns;
    constant Tsetup: time :=  15 ns;
    constant Thold : time :=   5 ns;
begin
    duv: muxed_ff port map(d0, d1, sel, q, qb,
                                clk, rst);

    clock_generator: clk <= not clk after cycle/2;

    test_procedure: process
        procedure sync_reset is
            ...
        end sync_reset;

        procedure load_d0(data: in std_logic) is
        begin
            rst <= '0';
            d0  <= data;
            d1  <= not data;
            sel <= '0';
            wait until clk = '1';
            wait for Thold;
            assert q == data and qb = not data;
            rst <= 'X';
            d0  <= 'X';
            d1  <= 'X';
            sel <= 'X';
            wait for cycle - Thold - Tsetup;
        end load_d0;

        procedure load_d1(data: in std_logic) is
            ...
        end load_d1;
    begin
        sync_reset;
        load_d0('1');
        ...
        assert FALSE severity FAILURE;
    end process test_procedure;
end test;
```

benches are divided into two major components: the reusable test harness, and the testcase-specific code.

Bus-functional models were assumed to be in the same process or module as the testbench.

In the previous chapter, stimulus generation and response checking were performed by abstracting operations using *procedures* or *tasks*. It was implied that these subprograms were implemented in the same *process* or *module* as the test sequence using them. Sample 6-1 shows where the task *load_d0*, first introduced in Sample 5-22, would have to be implemented in a Verilog testbench. Sample 6-2 shows the equivalent VHDL implementation.

Global access to signals declared at the module or architecture level was allowed.

With the suprograms located in the testbench module or process, they can be used by the *initial* and *always* blocks or processes implementing the testcase. The subprograms are simple to implement because the signals going to the device under verification can be driven directly through assignments to globally visible signals. Similarly, the outputs coming out of the device can be directly sampled as they too are globally visible.

The bus-functional models can be used by many testbenches.

All of the testbenches have to interface, through an instantiation, to the same design under verification. It is safe to assume that they all require the use of the same bus-functional models used to generate stimulus and to monitor response. These bus-functional models could be reused by all testbenches implemented for this design. If the interfaces being exercised or monitored by these bus-functional models have common interfaces found on other designs, they could even be reused by all testbenches for these other designs.

Use a low-level layer of reusable bus models.

Instead of a monolithic block, the testbenches should be structured with a low-level layer of reusable bus-functional models. This low-level layer is common to all testbenches for the design under verification and called the *test harness*. Each *testcase* would be implemented on top of these bus-functional models, as illustrated in Figure 6-1. The *testcase* and the *harness* together form a *testbench*.

Figure 6-1.
Structure of a testbench with reusable bus-functional models

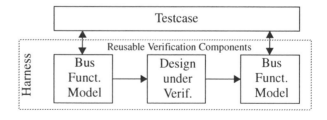

Insert reusable mid-level utility routines as required.	Many testbenches share some common functionality or need for interaction with the device under verification. Once the low-level features are verified, the repetitive nature of communicating with the device under verification can also be abstracted into higher-level utility routines. For example, low-level read and write operations to send and receive individual bytes can be encapsulated by utility routines to send and receive fixed-length packets. These, in turn, can be encapsulated in a higher-level utility routine to exchange variable-length messages with guaranteed error-free delivery.
The testcase can operate at the required level of abstraction.	A testcase verifying the low-level read and write operations would interface directly with the low-level bus-functional model, as shown in Figure 6-1. But once these basic operations are demonstrated to function properly, testbenches dealing with higher-level functions can use the higher-level utility routines, as shown in Figure 6-2.

Figure 6-2. Structure of a testbench with reusable utility routines

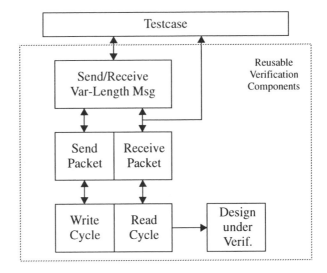

Procedural Interface

Define a procedural interface to the bus-functional model and utility routines.	For these verification components to be reusable by many testbenches, you must define a procedural interface independent of their detailed implementation. A procedural interface simply means that all the functionality of these components is accessed through

procedures or tasks, never through global variables or signals. It is similar to providing tasks or procedures to encapsulate operations. This gives flexibility in implementing or modifying the bus-functional models and utility routines without affecting the testcases that use them.

Provide flexibility through thin layers.

The verification components need to be flexible enough to provide the required functionality for all testbenches that use them. It is better to provide this flexibility by layering utility routines on top of general purpose lower-level routines and bus-functional models. This approach creates layers of procedural interfaces. The low-level layer provides detailed control whereas the higher-level provides greater abstraction. Do not attempt to implement all functionality in a single level. It would unduly complicate the implementation of the bus-functional models and increase the risk of introducing a functional failure.

Preserve the procedural interfaces.

By stimulating and monitoring a design through procedural interfaces, it removes the testcase from knowing the low-level details of the physical interfaces on the design. If the procedural interface is well-designed and can support different physical implementations, the physical interface of a design can be modified without having to modify any testbenches.

For example, a processor interface could be changed from a VME bus to a X86 bus. All that needs to be modified is the implementation of the CPU bus-functional model. If the procedural interface to the CPU bus-functional model is not modified, none of the testbenches need to be modified.

Another example would be changing a data transmission protocol from parallel to serial. As long as the testcases can still send bytes, they need not be aware of the change. Once you have defined a procedural interface, document it and hesitate to change it.

Development Process

Introduce flexibility as required.

When developing the low-level bus-functional models and the utility routines, do not attempt to write the ultimate verification component that includes every possible configuration option and operating mode. Use the verification plan to determine the functionality that is ultimately required. Architect the implementation of the verification component to provide this functionality, but implement incrementally. Start with the basic functions that are required by the basic testbenches. As testbenches progress toward exercising more complex functions, develop the required supporting functions by adding configurability to the bus-functional models or creating utility routines. As the verification infrastructure grows, the procedural interfaces are maintained to avoid breaking testbenches already completed.

Incremental development maximizes the verification efficiency.

This incremental approach minimizes your development effort: you won't develop functionality that turns out not to be needed. You also minimize your debugging effort, as you are building on functionality that has already been verified and debugged with actual testbenches. This approach also allows the development of the verification infrastructure to parallel the development of the testbenches, removing it from the critical path.

VERILOG IMPLEMENTATION

This section evolves an implementation of the test harness and testbench architecture. Starting with a monolithic testbench, the implementation is refined into layers of bus-functional models, utility packages, and testcases, with well-defined procedural interfaces. The goal is to obtain a flexible implemention strategy promoting the reusability of verification components. This strategy can be used in most Verilog-based verification projects.

Creating another testcase simply requires changing the control block.

In Sample 6-1, the entire testbench is implemented in a single level of hierarchy. If you were to write another testcase for the muxed flip-flop design using the same bus-functional models, you would have to replicate everything, except for the *initial* block that controls the testcase. The different testcases would be implemented by providing different control blocks. Everything else would remain the same. But replication is not reuse. It creates additional physical copies that have to be maintained. If you had to write fifty testbenches, you would have to maintain fifty copies of the same bus-functional models.

Move the testcase control into a higher-level of hierarchy.

The implementation of reusable verification components in Verilog is relatively simple. Leave the portions of the testbench that are the same for all testbenches in the level of hierarchy immediately surrounding the design under verification and move the control structure unique to each testcase into a higher level of hierarchy.

Instead of invoking the bus-functional models directly, they are invoked using a hierarchical name. The level of hierarchy containing the reusable verification components, called the *test harness*, provides the procedural interface to the design under verification.

Figure 6-3 illustrates the difference in testbench structure. Figure 6-3(a) shows the original, non-reusable structure, while Figure 6-3(b) shows the same testbench using the reusable test harness structure. Sample 6-3 shows the implementation of the testcase shown earlier in Sample 6-1, using a reusable test harness. Notice how the tasks are now invoked using a hierarchical name.

Figure 6-3. Verilog implementation of a testbench using a test harness structure

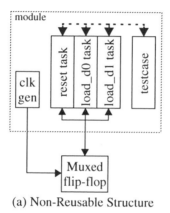

(a) Non-Reusable Structure

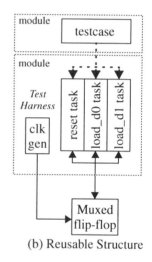

(b) Reusable Structure

The test harness includes everything needed to operate the design.

The test harness should be self-contained and provide all signals necessary to properly operate the design under verification. In addition to all the low-level bus-functional models, it should include the clock and reset generators. The reset generator should be encapsulated in a task. This lets testcases trigger the reset operation at will, if required.

Packaging Bus-Functional Models

Bus-functional models can be reused between harnesses.

The structure shown in Figure 6-3 lets the test harness be reused between many testcases on the same design under verification. But it does not help the reusability of bus-functional models between test harnesses for different designs.

```
module testcase;

harness th();

initial
begin: test_sequence
   th.sync_reset;
   th.load_d0(1'b1);
   ...
   $finish;
end
endmodule

module harness;

reg rst, d0, d1, sel, clk;
wire q, qb;

muxed_ff duv(d0, d1, sel, q, qb, clk, rst);

parameter cycle = 100,
          Tsetup = 15,
          Thold  =  5;

always
begin
   #(cycle/2) clk = 1'b0;
   #(cycle/2)     = 1'b1;
end

task sync_reset;
   ...
endtask

task load_d0;
   input data;
begin
   ...
end
endtask

task load_d1;
   ...
endtask
endmodule
```

All the tasks providing a complete bus-functional model for a given interface should be packaged to make them easy to reuse between test harnesses. For example, all of the tasks encapsulating the operations of a PCI bus should be packaged into a single PCI bus-functional model package to facilitate their reuse. Some of these tasks have been shown in Sample 5-61 and Sample 5-63.

Package a bus-functional model in its own level of hierarchy.

We made the test harness reusable by isolating it from its testcases in its own level of hierarchy. The testcases using the test harness simply have to instantiate it and use its procedural interface through hierarchical calls to use it. A similar strategy can be used to make bus-functional models reusable. All of the tasks encapsulating the operations are located in a module, creating a self-contained bus-functional model. For more details, see "Encapsulating Useful Subprograms" on page 94.

The procedural interface is accessed hierarchically and the physical interface is accessed through pins.

The signals driven or monitored by the tasks are passed to the bus-functional model through pins. The bus-functional model is instantiated in the test harness and its pins are properly connected to the design under verification. The tasks in the bus-functional model provide its procedural interface. They are called by the testcase using hierarchical names through the test harness.

Figure 6-4.
Test harness structures using an i386SX bus-functional model

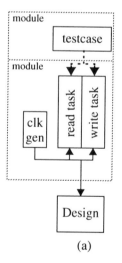 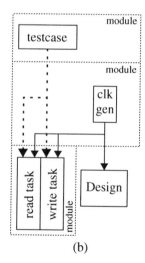

(a) (b)

Figure 6-4(a) shows the structure of a test harness with a non-reusable bus-functional model of an Intel 386SX processor. The bus-functional model is composed of two tasks: *read* and *write*

Figure 6-4(b) shows a functionally equivalent harness using a properly packaged bus-functional model. Sample 6-4 shows the skeleton Verilog code implementing an Intel 386SX bus-functional model, while Sample 6-5 and Sample 6-6 show how it can be used by the test harness and testcase, respectively.

Sample 6-4.
Packaged bus-functional model for a i386SX

```
module i386sx(clk, addr, ads, rw, ready, data);
input          clk;
output [23:0]  addr;
output         ads;
output         rw;
input          ready;
inout  [15:0]  data;

reg [23:0] addr;
...
reg [15:0] data_o;
assign data = data_o;

initial
begin
    ads     = 1'b1;
    data_o  = 16'hZZZZ;
end

task read;
    ...
endtask

task write;
    ...
endtask
endmodule
```

Utility Packages

Mid-level utility routines are packaged in separate modules.

The utility routines that provide additional levels of abstraction to the testcases are also composed of a series of *tasks* and *functions*. They can be encapsulated in separate modules, using hierarchical names to access the lower-level procedural interfaces. The utility routines they provide would also be called using a hierarchical name.

Utility packages are never instantiated.

Because there is no *wire* or *register* connectivity involved between a utility package and the lower-level procedural interfaces, they need not be instantiated. They form additional simulation top-level

Sample 6-5.
Test harness using the packaged i386SX bus-functional model

```
module harness;

reg        clk;
wire [23:0] addr;
...
wire [15:0] data;

i386sx cpu(clk, addr, ads, rw, ready, data);
design dut(clk, addr, ads, rw, ready, data, ...);

always
begin
    #50 clk = 1'b0;
    #50 clk = 1'b1;
end

task reset
begin
    disable cpu.read;
    disable cpu.write;
    ...
end
endtask

endmodule
```

Sample 6-6.
Testcase using the packaged i386SX bus-functional model

```
module testcase;

harness th();

initial
begin: test_procedure
    reg [15:0] val;

    th.reset;
    th.cpu.read(24'h00_FFFF, val);
    val[0] = 1'b1;
    th.cpu.write(24'h00_FFFF, val);
    ...
end
endmodule
```

modules, running in parallel with the testbench and the design under verification. They can access the tasks and functions in the test harness using *absolute* hierarchical names.

Their own functions and tasks are also called using absolute hierarchical names. Sample 6-7 shows the implementation of a simple utility routine to send a fixed-length 64-byte packet, 16 bits at a time, via the i386SX bus using the Intel 386SX bus-functional model shown in Sample 6-4 and used in the test harness shown in Sample 6-5. Notice how an absolute hierarchical name is used to access the *write* task in the CPU bus-functional model in the harness in the testcase.

Sample 6-7.
Utility package on test harness using packaged i386SX bus-functional model

```
module packet;

task send;
   input [64*8:1] pkt;
   reg    [  15:0] word;
   integer i;
begin
   for (i = 0; i < 32; i = i + 1) begin
      word = pkt[16:1];
      testcase.th.cpu.write(24'h10_0000 + i,
                                     word);
      pkt = pkt >> 16;
   end
end
endmodule
```

The harness is not intantiated either.

If the test harness is instantiated by the top-level testcase module, as shown in Sample 6-6, the name of the testcase module is part of any absolute hierarchical name. You can standardize on using a single predefined module name for all testcase modules and restrict them to a single level of hierarchy, with the test harness instantiated under a predefined instance name.

A better alternative is to leave the test harness uninstantiated, forming its own simulation top-level module. The testcases would simply use absolute hierarchical names instead of the relative hierarchical names to access tasks and functions in the test harness.

Sample 6-8 shows the testcase previously shown in Sample 6-6, but using absolute hierarchical names into an uninstantiated test harness. It also uses the packaged utility routine modified in Sample 6-9 to use the uninstantiated test harness. Figure 6-5 shows the structure of the simulation model, with the multiple top-levels.

Sample 6-8.
Testcase using uninstantiated test harness

```
module testcase;

initial
begin: test_procedure
    reg [15:0] val;
    reg [64:8:1] msg;

    harness.reset;
    harness.cpu.read(24'h00_FFFF, val);
    val[0] = 1'b1;
    harness.cpu.write(24'h00_FFFF, val);
    ...
    packet.send(msg);
end
endmodule
```

Sample 6-9.
Utility package on uninstantiated test harness

```
module packet;

task send;
    input [64*8:1] pkt;
    reg [15:0] word;
    integer i;
begin
    for (i = 0; i < 32; i = i + 1) begin
        word = pkt[16:1];
        harness.cpu.write(24'h10_0000 + i, word);
        pkt = pkt >> 16;
    end
end
endmodule
```

Figure 6-5.
Simulation structure with uninstantiated harness and utility package

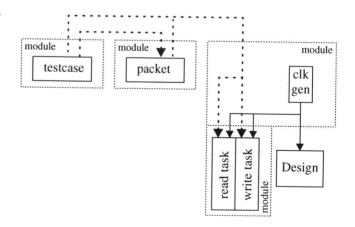

Additional top-
levels are added to
the command line.
It is very easy to create a Verilog simulation with multiple top-level
modules. They are included in the simulation by simply adding
their filename or module name to the simulation command. If you
are using a simulator that compiles and elaborates the simulation
structure in a single command, such as *Verilog-XL* or *VCS*, simply
specify the additional filenames that compose the other top-levels.
Assuming that the files involved in creating the structure shown in
Figure 6-5, are named *testcase.v, packet.v, harness.v, i386sx.v*, and
design.v, the command to use with *Verilog-XL* to simulate them
would be:

```
% verilog testcase.v packet.v harness.v \
        i386sx.v design.v
```

For a simulation tool with separate compilation and elaboration
phases, such as *ModelSim*, all of the required top-level modules
must be identified to the simulation command:

```
% vlog testcase.v packet.v harness.v \
        i386sx.v design.v
% vsim testcase packet harness
```

As shown in Sample 6-10, the simulator displays the names of all
the top-level modules in the simulation, and simulates them seam-
lessly, as one big model.

Sample 6-10.
Simulator dis-
playing top-
level modules

```
. . .
The top-level modules are:
harness
testcase
packet

Simulation begins...
. . .
```

Instantiating utility
packages is too
restrictive.
You might be tempted to require that all packages be instantiated
one on top of each other. Lower-level utility package would instantiate the test harness, and higher-level packages would instantiate
the lower-level packages. The structure of the testbench would thus
follow the structure of the packages and only relative hierarchical
names would be used. Unfortunately, the reverse would be occur-
ing: the packages would be forced into following the structure of
the testbench.

Follow a logical structure.

Requiring that every utility package be instantiated restricts their structure to a single hierarchical line. The test harness encapsulates the design and its surrounding bus-functional models into a single module. From that point on, only one module can be layered on top of it. It is not possible to create a tree in the hierarchy where all the branches terminate in the test harness.

Figure 6-6(a) shows the only possible structure allowed by instantiating all packages according to their abstraction layer. It is impossible to create the logical structure shown in Figure 6-6(b). The latter can be implemented using uninstantiated packages and absolute hierarchical names.

Figure 6-6.
Simulation structures with harness and utility packages

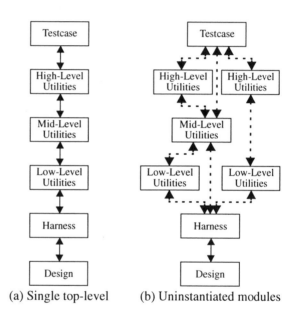

(a) Single top-level   (b) Uninstantiated modules

Avoid cross-references in utility routines.

Because the utility packages are implemented in uninstantiated modules, they create a flat structure of globally visible procedural interfaces. The models do not enforce that they are used in a strictly layered fashion. It is possible - and tempting - to write utility packages that cross-reference themselves.

Sample 6-11 illustrates an example of cross-references, where packages use routines from each other. Cross-references make two packages inter-dependent. It is not possible to debug and verify one separately from the other. It also makes the packages more difficult to reuse as they might have to be decoupled to be fitted into a different simulation environment.

When designing utility packages, stick to a strict set of layers. Packages can access utility routines in lower layers or within themselves, but never in a sibling package at the same level of abstraction. If a need for cross-references arises, question your design of the package set, or consider merging both packages into a single one.

Sample 6-11.
Packages with
cross-refer-
ences

```
module syslog;

task note;
    input [80*8:1] msg;

    $write("NOTE: %0s\n", msg);
endtask

task terminate;
begin
    $write("Simulation terminated normally\n";
    watchdog.shutdown;
    $finish;
end
endtask

endmodule

module watchdog;
task shutdown;
begin
    syslog.note("Watchdog shutting down...");
    ...
end
endtask
endmodule
```

VHDL IMPLEMENTATION

This section evolves an implementation of the test harness and testbench architecture. Starting with a monolithic testbench, the implementation is refined into client/server bus-functional models, access and utility packages, and testcases. The goal is to obtain a flexible implemention strategy promoting the reusability of verification components. This strategy can be used in most VHDL-based verification projects.

Creating another testcase simply requires changing the control block.

In Sample 6-2, the entire testbench is implemented in a single level of hierarchy. If you were to write another testcase for the muxed flip-flop design using the same bus-functional models, you would have to replicate everything, except for the body of the *process* that controls the testcase. The different testcases would be implemented by providing different sequential statements. Everything else would remain the same, including the procedures in the process declarative region. But replication is not reuse. It creates additional physical copies that have to be maintained. If you had to write fifty testbenches, you would have to maintain fifty copies of the same bus-functional models.

Packaging Bus-Functional Procedures

Bus-functional models can be located in a *package* to be reused.

One of the first steps to reducing the maintenance requirement is to move the bus-functional procedures from the process declarative regions to a *package*. These procedures can be used outside of the *package* by each testbench that requires them.

Sample 6-12.
Bus-functional procedures for an i386SX

```
package i386sx is

subtype add_typ is std_logic_vector(23 downto 0);
subtype dat_typ is std_logic_vector(15 downto 0);

procedure read(        raddr: in      add_typ;
                       rdata: out     dat_typ;
                signal clk  : in      std_logic;
                signal addr : out     add_type;
                signal ads  : out     std_logic;
                signal rw   : out     std_logic;
                signal ready: in      std_logic;
                signal data : inout   dat_typ);

procedure write(       waddr: in      add_typ;
                       wdata: in      dat_typ;
                signal clk  : in      std_logic;
                signal addr : out     add_type;
                signal ads  : out     std_logic;
                signal rw   : out     std_logic;
                signal ready: in      std_logic;
                signal data : inout   dat_typ);
end i386sx;
```

They require *signal*-class arguments.

However, bus-functional procedures, once moved into a *package*, require that all driven and monitored signals be passed as *signal*-class arguments (see "Encapsulating Bus-Functional Models" on page 97 and Sample 4-16 on page 98). Sample 6-12 shows the *package* declaration of bus-functional model procedures for the Intel 386SX processor. Notice how all the signals for the processor bus are required as *signal*-class arguments in each procedure.

Bus-functional model procedures are cumbersome to use.

Sample 6-13 shows a process using the procedures declared in the package shown in Sample 6-12. They are very cumbersome to use as all the signals involved in the transaction must be passed to the bus-functional procedure. Furthermore, there would still be a lot of duplication across multiple testbenches. Each would have to declare all interface signals, instantiate the component for the design under verification, and properly connect the ports of the component to the interface signals. With today's ASIC and FPGA packages, the number of interface signals that need to be declared, then mapped, can easily number in the hundreds. If the interface of the design were to change, even minimally, all testbenches would need to be modified.

Sample 6-13.
Using bus-
functional pro-
cedures

```
use work.i386sx.all;
architecture test of bench is
    signal clk  : std_logic;
    signal addr : add_type;
    signal ads  : std_logic;
    signal rw   : std_logic;
    signal ready: std_logic;
    signal data : dat_typ;
begin

    duv: design port map (..., clk, addr, ads,
                               rw, ready, data, ...);

    testcase: process
        variable data: dat_typ;
    begin
        ...
        read(some_address, data,
            clk, addr, ads, rw, ready, data);
        ...
        write(some_other_address, some_data,
            clk, addr, ads, rw, ready, data);
        ...
    end process testcase;
end test;
```

Creating a Test Harness

The test harness contains declarations and functionality common to all testbenches.

To reduce the amount of duplicated information from testbench to testbench, you must factor out their common elements into a single structure that they will share. The common elements in all testbenches for a single design are:

• Declaration of the component

• Declaration of the interface signals

• Instantiation of the design under verification

• Mapping of interface signals to the ports of the design

• Mapping of interface signals to the signal-class arguments of bus-functional procedures.

Figure 6-7.
VHDL implementation of a testbench using a test harness structure

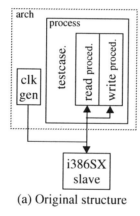

(a) Original structure

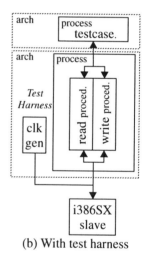

(b) With test harness

Use an intermediate level of hierarchy to encapsulate the test harness.

Figure 6-7(a) illustrates the structure of the testbench shown in Sample 6-13. The *read* and *write* procedures are shown in their invocation context, not their declaration context. Whether you use bus-functional procedures declared in the process declarative region, or in a package, the testbench structure remains the same. This is because the signal drivers are associated with the calling process.

Figure 6-7(b) illustrates how the same testbench can be structured to take advantage of the replicated functionality in multiple testbenches. The replicated functionality is located in a lower-level architecture, called a *test harness*. The testcase drives the design under verification by instructing *processes* in the test harness to perform various operations.

A process in the harness owns the bus-functional procedures.

Because the signals that interface to the design under verification are local to the test harness, the bus-functional procedures must be called by a local process. The bus-functional procedures use the drivers associated with that local process. The testcase control process must instruct the local process, through control signals, to perform the appropriate cycle and return any relevant information.

This local process is often call a *server* process, while the testbench control process is called a *client* process. The control signals have to be visible to both the client and server processes, located in a different architecture. This can be accomplished in two ways:

- Passing them as ports on the test harness entity

- Making them global signals in a package.

Sample 6-14.
Client/server control package

```
package i386sx is

type do_typ is (read, write);

subtype add_typ is std_logic_vector(15 downto 0);
subtype dat_typ is std_logic_vector(15 downto 0);

type to_srv_typ is record
  do  : do_typ;
  addr: add_type;
  data: dat_typ;
end record;

type frm_srv_typ is record
   data: dat_typ;
end record;

signal to_srv : to_srv_typ;
signal frm_srv: frm_srv_typ;

end i386sx;
```

Since a package is required to contain their type definitions, the latter does not require additional library units. Furthermore, using global signals eliminates the need for each testbench architecture to declare them, then mapping them to the ports of the test harness. Sample 6-14 shows an implementation of the client/server control package for controlling the i386SX bus-functional procedures. The server process is in the test harness shown in Sample 6-15.

```
use work.i386sx.all;
architecture test of bench is
    signal clk  : std_logic;
    signal addr : add_type;
    signal ads  : std_logic;
    signal rw   : std_logic;
    signal ready: std_logic;
    signal data : dat_typ);
begin

    duv: design port map (..., clk, addr, ads,
                               rw, ready, data, ...);

    i386sx_server: process
        variable data: dat_typ;
    begin
        wait on to_srv'transaction;
        if to_srv.do = read then
            read(to_srv.addr, data,
                clk, addr, ads, rw, ready, data);
        elsif to_srv.do = write then
            write(to_srv.addr, to_srv.data,
                clk, addr, ads, rw, ready, data);
        end if;
        frm_srv.data <= data;
    end process i386sx_server;
end test;
```

Use 'transaction to synchronize operations.

Notice how the server process is sensitive to transactions on the to_srv control signal. This way, it is triggered after every assignment to the control signal, whether they are the same or not. Had the process been sensitive to events on to_srv, the second of two consecutive identical operations, as shown in Sample 6-16, would be missed.

Records are used to implement the control signals.

User-defined record types are used for the client/server control signals. Even if the record contains a single element, such as the frm_srv_typ record in Sample 6-14. A record is used to minimize maintenance if the control protocol between the client and the server needs to be modified. Fields can be removed, added or modified without affecting the type declaration of the control signals

```
to_srv <= (do   => write,
             addr => (others => '1'),
             data => (others => '0'));
wait on frm_srv.data'transaction;
to_srv <= (do   => write,
             addr => (others => '1'),
             data => (others => '0'));
```

themselves, minimizing the impact on clients and server processes using them.

Abstracting the Client/Server Protocol

The client must
properly operate
the control signals
to the server pro-
cess.

Sample 6-17 shows a client process accessing the services provided by the i386SX server process in the test harness shown in Sample 6-15. Notice how the client process waits for a transaction on the return signal to detect the end of the operation. This behavior detects the end of an operation that produces the same result as the previous one. If the client process had been sensitive to events on the return *frm_srv* signal, the end of the operation could have been detected only if it produced a different result from the previous one.

```
use work.i386sx.all;
architecture test of bench is
begin

    i386sx_client: process
        variable data: dat_typ;
    begin
        . . .
        -- Perform a read
        to_srv.do   <= read;
        to_srv.addr <= ...;
        wait on frm_srv'transaction;
        data := frm_srv.data;
        . . .
        -- Perform a write
        to_srv.do   <= write;
        to_srv.addr <= ...;
        to_srv.data <= ...;
        wait on frm_srv'transaction;
        . . .
    end process i386sx_client;
end test;
```

Encapsulate the client/server operations in procedures.

Defining a communication procotol on signals between the client and the server processes does not seem to accomplish anything. Instead of having to deal with a physical interface documented in the design specification, we have to deal with an arbitrary protocol with no specification. Just as the operation on the physical interface can be encapsulated, the operations between the client and server can also be encapsulated in procedures. This encapsulation removes the client process from knowing the details of the protocol with the server. The protocol can be modified without affecting the testcases using it through the procedures encapsulating the operations.

Sample 6-18.
Client/server access package

```
package i386sx is

type do_typ is (read, write);

subtype add_typ is std_logic_vector(15 downto 0);
subtype dat_typ is std_logic_vector(15 downto 0);

type to_srv_typ is record
  do  : do_typ;
  addr: add_type;
  data: dat_typ;
end record;

type frm_srv is record
  data: dat_typ;
end record;

procedure read(        addr  : in  add_typ;
                       data  : out dat_typ;
               signal to_srv : out to_srv_typ;
               signal frm_srv: in  frm_srv_typ);

procedure write(       addr  : in  add_typ;
                       data  : in  dat_typ;
               signal to_srv : out to_srv_typ;
               signal frm_srv: in  frm_srv_typ);

signal to_srv : to_srv_typ;
signal frm_srv: frm_srv_typ;

end i386sx;
```

Put the server access procedure in the control package.

The server access procedures should be located in the package containing the type definition and signal declarations. Their implementation is closely tied to these control signals and should be located with them. Sample 6-18 shows how the *read* and *write* access procedures would be added to the package previously shown in Sample 6-14.

Client processes
use the server
access procedures.

The client processes are now free from knowing the details of the protocol between the client and the server. To perform an operation, they simply need to use the appropriate access procedure. The pair of control signals to and from the server must be passed to the access procedure to be properly driven and monitored. Sample 6-19 shows how the client process, originally shown in Sample 6-17, is now oblivious to the client/server procotol.

Sample 6-19.
Client process
using server
access proce-
dures

```
use work.i386sx.all;
architecture test of bench is
begin

    i386sx_client: process
        variable data: dat_typ;
    begin
        ...
        -- Perform a read
        read(..., data, to_srv, frm_srv);
        ...
        -- Perform a write
        write(..., ..., to_srv, frm_srv);
        ...
    end process i386sx_client;
end test;
```

The testcase must still pass signals to and from the bus-functional access procedures. So, what has been gained from the starting point shown in Sample 6-13? The answer is: a lot.

Testbenches are
now removed from
the physical
details.

First, the testcase need not declare all of the interface signals to the design under verification, nor instantiate and connect the design. These signals can number in the high hundreds, so a significant amount of work duplication has been eliminated.

Second, no matter how many signals are involved in the physical interface, you need only pass two signals to the bus-functional access procedures. The testcases are completely removed from the physical interface of the design under verification. Pins can be added or removed and polarities can be modified without affecting the existing testcases.

Managing Control Signals

Use separate unre-
solved *to* and *from*
control signals.

Using two control signals, one to send control information and syn-chronization to the server, and vice-versa is the simplest solution. The alternative is to use a single signal, where both client and server processes each have a driver. A resolution function would be required, including a mechanism for differentiating between the value driven from the server and the one driven from the client.

You could use a
single resolved
control signal.

If you want to simplify the usage of the access procedures and the syntax of the client processes, a single resolved control signal can be used between the client and server processes. Instead of having to pass two signals to every server access procedures, only one signal needs to be passed. The price is additional development effort for the server access package - but since it is done only once for the entire design, it may be worth it.

But the risks out-
weigh the benefits.

Inserting a resolution function between the client and the server also introduces an additional level of complexity. It can make debugging the client/server protocol and the testcases that use it more tedious. It also makes it possible to have separate processes drive the control signal to the server process. Because that signal is now resolved, no error would be generated because of the multiple driver. Without proper interlocking of the parallel requests to the server, this would create a situation similar to Verilog's non-reen-trant tasks.

Use qualified
names for access
procedures.

In a test harness for a real design, there may be a dozen server pro-cesses, each with their own access package and procedures. A real-life client process, creating a complex testcase, uses all of them. It may be difficult to ensure that all identifiers are unique across all access packages. In fact, making identifier uniqueness a require-ment would place an undue burden on the authoring and reusability of these packages.

The identifier collision problem can be eliminated by using quali-fied names when using access procedures. Sample 6-20 shows a testbench using qualified names to access the *read* procedure out of the *i386sx* package. Notice how the *use* statement for the package does not specify "*.all*" to make all of the identifiers it contains visi-ble.

Sample 6-20.
Client process
using quali-
fied identifiers

```
use work.i386sx;
architecture test of bench is
begin
    i386sx_client: process
        variable data: i386sx.dat_typ;
    begin
        ...
        -- Perform a read
        i386sx.read(..., data, i386sx.to_srv,
                         i386sx.frm_srv);
        ...
    end i386sx_client;
end test;
```

Multiple Server Instances

Provide an array of
control signals for
multiple instances
of the same server
processes.

Designs often have multiple instances of identical interfaces. For example, a packet switch design would have multiple packet input and output ports, all using the same physical protocol. Each can be stimulated or monitored using separate server processes using the same bus-functional procedures. The clients needs to have a way to identify which server process instance they want to operate on to perform operations on the proper interface on the design.

Using an array of control signals, one pair for each server, meets this requirement. Sample 6-21 shows the access package containing an array of control signals, while Sample 6-22 shows one instance of a server process.

Sample 6-21.
Array of cli-
ent/server con-
trol signals for
multiple serv-
ers

```
package i386sx is
...

type to_srv_ary_typ  is array(integer range <>)
    of to_srv_typ;
type frm_srv_ary_typ is array(integer range <>)
    of frm_srv_typ;

signal to_srv  : to_srv_ary_typ (0 to 7);
signal frm_srv : frm_srv_ary_typ(0 to 7);

end i386sx;
```

Sample 6-22.
One instance of server process using an array of control signals

```
use work.i386sx.all;
architecture test of bench is
    ...
begin

    ...

    i386sx_server: process
        variable data: dat_typ;
    begin
        wait on to_srv(3)'transaction;
        if to_srv(3).do = read then
            read(to_srv(3).addr, data, ...);
        elsif to_srv(3).do = write then
            write(to_srv(3).addr, to_srv(3).wdat,
                    ...);
        end if;
        frm_srv(3).rdat <= (others => 'X');
    end process i386sx_server;
end test;
```

You may be able to use the for-generate statement.

If the physical signals for the multiple instances of a port are properly declared using arrays, a *for-generate* statement can be used to automatically replicate the server process. Sample 6-23 illustrates this.

Sample 6-23.
Generating multiple instances of a server process

```
use work.rs232.all;
architecture test of bench is
    signal tx: std_logic_vector(0 to 7);
    ...
begin

    duv: design port map(tx0 => tx(0),
                         tx1 => tx(1), ...);

    servers: for I in tx'range generate
        process
            variable data: dat_typ;
        begin
            wait on to_srv(I)'transaction;
            receive(data, tx(I));
            frm_srv(I).rdat <= data;
        end process;
    end generate servers;
end test;
```

Testbench genera-
tion tools can help
in creating the test
harness and access
packages.

If you are discouraged by the amount of work required to implement a VHDL test harness and access packages, remember that it
will be the most leveraged verification code. It will be used by all
testbenches so investing in implementing a test harness that is easy
to use returns the additional effort many times. Testbench generation tools, such as *Quickbench* by Chronology, can automate the
generation of the test harness from a graphical specification of the
timing diagrams describing the interface operations.

Utility Packages

Utility routines are
packaged in sepa-
rately.

The utility routines that provide additional levels of abstraction to
the testcases are also composed of a series of *procedures*. They can
be encapsulated in separate packages using the lower-level access
packages. Sample 6-24 shows the implementation of a simple utility routine to send a fixed-length 64-byte packet, 16 bits at a time,
via thei386SX bus using the Intel 386SX access package shown in
Sample 6-18. Sample 6-25 shows a testcase using the utility package defined in Sample 6-24.

Sample 6-24.
Utility pack-
age using
i386SX bus-
functional
model access
package

```
use work.i386sx.all;
package packet is

type packet_typ is array(integer range <>)
   of std_logic_vector(15 downto 0);

procedure send(       pkt    : in  packet_typ;
                signal to_srv : out to_srv_typ;
                signal frm_srv: in  frm_srv_typ);

end packet;

package body packet is

procedure send(       pkt    : in  packet_typ;
                signal to_srv : out to_srv_typ;
                 signal frm_srv: in  frm_srv_typ)
   is
   begin
      for I in pkt'range loop
         write(..., pkt(I), to_srv, frm_srv);
      end loop;
   end send;
end packet;
```

<table>
<tr><td>

Sample 6-25.
Testcase using
utility proce-
dure

</td><td>

```
use work.i386sx;
use work.packet;
architecture test of bench is
begin

    testcase: process
        variable pkt: packet_typ(0 to 31);
    begin
        . . .
        -- Send a packet on i386 i/f
        packet.send(pkt, i386sx.to_srv,
                         i386sx.frm_srv);
        . . .
    end process testcase;
end test;
```

</td></tr>
</table>

AUTONOMOUS GENERATION AND MONITORING

This section explains how properly packaged bus-functional models can become active entities. They can remove the testcase from the tedious task of generating background or random data, or performing detailed response checking.

Packaged bus-functional models create an opportunity.

Once the bus-functional procedures are moved in a module or controlled by an entity/architecture independent from the testcase, it creates an opportunity to move the tedious housekeeping tasks associated with using these bus-functional models along with them.

The packaged bus-functional models can now contain *processes* and *always* or *initial* blocks. These concurrent behavioral descriptions can perform a variety of tasks such as safety checks, data generation, or collecting responses for later retrieval. Instead of requiring constant attention by the testcase control process, these packaged bus-functional models could instead be configured to autonomously generate data according to configurable parameters. They could also be configured to monitor output response, looking for unusual patterns or specific data, and notifying the testbenches only when exceptions occur.

Autonomous Stimulus

Protocols may require more data than is relevant for the testcase.

Imagine a protocol on a physical interface that requires data to be continuously flowing. This procotol would obviously include a "data not valid" indication.

Figure 6-8 shows such an interface, where ATM cells are constantly flowing, aligned with a cell boundary marker. If this interface were to be stimulated using procedures only, the testcase would have to implement a control structure to continuously call the *send_cell* procedure not to violate the protocol. Most of the time, an invalid or predefined cell would be sent. But under further control by the testcase, a valid cell, relevant to the testcase, would be appropriately inserted. This control structure will likely have to be repeated in all testbenches generating input for that interface.

Figure 6-8.
ATM protocol
requiring
continuous
data flow

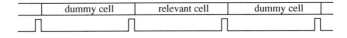

The access procedures would interface with a transmission process.

The *send_cell* procedure, contrary to previous implementations, would not immediately cause a cell to be sent on the physical interface and return once it has been transmitted. Instead, it would synchronize with the process transmitting dummy cells to have the relevant cell inserted in the data stream at the appropriate point.

It could also provide blocking or non-blocking implementations. In a blocking implementation, the *send_cell* procedure would return only when the cell was transmitted. In a non-blocking implementation, the *send_cell* procedure would return immediately, queueing the cell for future transmission.

Sample 6-26 shows an implementation of a blocking *send_cell* procedure in Verilog, while Sample 6-27 and Sample 6-28 show a non-blocking implementation in VHDL. Both blocking or non-blocking implementations could be provided and selected using an additional argument to the procedure.

Sample 6-26.
Blocking access procedure in bus-functional model

```
module atm_src(...);
...

task xmit_cell;
    input [...] cell;
begin
    ...
end
endtask

reg blocked;
task send_cell;
    input [...] cell;
begin
    blocked = 1'b1;
    wait blocked === 1'b0;
end
endtask

reg [...] dummy_cell;
always
begin
    if (blocked === 1'b1) begin
        xmit_cell(send_cell.cell);
        blocked = 1'b0;
    end else begin
        xmit_cell(dummy_cell);
    end
end
endmodule
```

Sample 6-27.
Non-blocking access procedure

```
package body atm_gen is

procedure send_cell(       cell: in   atm_cell;
                    signal to_srv : out to_srv_typ;
                    signal frm_srv: in   frm_srv_typ)
is
    to_srv.cell <= cell;
    wait on frm_srv'transaction;
end send_cell;

end atm_gen;
```

Sample 6-28.
Server pro-
cesses sup-
porting non-
blocking
access proce-
dure

```
use work.atm_gen.all;
architecture test of bench is
    subtype queue_idx is integer range 0 to 99;
    type atm_cell_array is array(queue_idx)
        of atm_cell;
    signal cell_queue: atm_cell_array;
    signal tail       : queue_idx;
begin

    non_block_srv: process
    begin
        wait on to_srv'transaction;
        cell_queue(tail) <= to_srv.cell;
        tail <= (tail + 1) rem cell_queue'length;
        frm_srv.done <= not frm_srv.done;
    end process non_block_srv;

    apply_cell: process
        variable head: queue_idx;
    begin
        if head = tail then
            wait until head /= tail;
        end if;
        ...
        head := (head + 1) rem cell_queue'length;
    end process apply_cell;

end test;
```

Random Stimulus

The content of
generated data can
be random.

It is a small step from automatically repeating the same stimulus to generating random stimulus. Instead of applying invalid or pre-defined data values or sequences, the packaged bus model can contain an algorithm to generate random data, in a random sequence, at random intervals.

Sample 6-29 shows the implementation of the i386SX bus-functional model from Sample 6-4 modified to randomly generate read and write cycles at random intervals within a specified address range. The *always* block could be easily modified to be interrupted by requests from the testcase to issue a specific read or write cycle on demand.

Sample 6-29.
Bus-func-
tional model
for a i386SX
generating
random cycles

```
module i386sx(...);

task read;
   ...
endtask

task write;
   ...
endtask

always
begin: random_generator
   reg [23:0] addr;
   reg [15:0] data;

   // Random interval (0-255)
   #($random >> 24);

   // Random even address
   addr[23:21] = 3'b000;
   addr[20: 1] = $random;
   addr[    0] = 1'b0;

   // Random read or write
   if ($random % 2) begin
      // Write random data
      write(addr, $random);
   end else begin
      // Read from random address
      read(addr, data);
   end
end
endmodule
```

Autonomous gen-
erators can help
compute the
expected response.

If the autonomous generators are given enough information, they may be able to help in the output verification. For example, the strategy used to verify the packet router illustrated in Figure 5-31 requires that a description of the destination be written in the payload of each packet using the format illustrated in Figure 5-32. The header and filler information could be randomly generated, but the description of the expected destination is a function of the randomly generated header.

Similarly, the CRC is computed based on the randomly generated payload and the destination descriptor. Sample 6-30 shows how such a generator could be implemented. The procedural interface (not shown) would be used to start and stop the generator, as well as filling the routing information in *rt_table*.

Sample 6-30.
Random generator helping to verify output

```
always
begin: monitor
   reg `packet_typ pkt;

   // Generate the header
   pkt`src_addr = my_id;
   pkt`dst_addr = $random;
   // Which port does this packet goes to?
   pkt`out_port_id = rt_table[pkt`dst_addr];
   // Next in a random stream
   pkt`strm_id = {$random, my_id};
   pkt`seq_num = seq_num[pkt`strm_id];
   // Fill the payload
   pkt`filler = $random;
   pkt`crc = computer_crc(pkt);
   // Send the packet
   send_pkt(pkt);

   seq_num[pkt`strm_id] =
      seq_num[pkt`strm_id] + 1;
end
```

Injecting Errors

Generators can be configured to inject errors.

If the design is supposed to detect errors in the interface protocol, you must be able to generate errors to ensure that they are properly detected. An autonomous stimulus generator could randomly generate errors. It could also have a procedural interface to inject specific errors at specific points in time. Sample 6-31 shows the generator from Sample 6-30 modified to corrupt the CRC on one percent of the packets. To make sure they are properly dropped, the sequence number is not incremented for corrupted packets.

Autonomous Monitoring

Monitors must always be listening.

In Chapter 5, we differentiated between a generator and a monitor based on who is in control of the timing. If the testbench controls when and if an operation occurs, then a generator is used. If the design under verification controls the timing and if an operation

Sample 6-31.
Randomly
injecting
errors

```
always
begin: monitor
    pkt'crc = computer_crc(pkt);

    ...
    // Randomly corrupt the CRC for 1% of cells
    if ($random %100 == 0) begin
        pkt'seq_num = $random;
        pkt'crc = pkt'crc ^ (1<<($random % 8));
    end else begin
        seq_num[pkt'strm_id] =
            seq_num[pkt'strm_id] + 1;
    end

    // Send the packet
    send_pkt(pkt);
end
```

occurs, then a monitor is used. In the latter situation, the testbench must be continuously listening to potential operations. Otherwise, activity on the output signals can be missed.

Figure 6-9 illustrates the timing of the behavior required by the testbench. The monitoring bus-functional procedures must be continuously called, with zero-delay between the time they return and the time they are called back. If there are gaps between invocation of the monitoring procedures, as shown in Figure 6-10, some valid output may be missed.

Figure 6-9.
Proper call
timing for
monitoring
procedures

Figure 6-10.
Delays
between calls
to monitoring
procedures

Usage errors can
be detected.

The role of a testbench is to detect as many errors as possible. Even from within the testbench itself. As the author of the test harness, your responsibility is not only to detect errors coming from the design under verification, but also to detect errors coming from the testcase. If an unexpected output operation goes unnoticed because the testcase was busy checking the response from the previous cycle, it creates a crack that a bad design can slip through.

An autonomous output monitor can verify that the monitoring procedures are properly used. If activity is noticed on the output interface and the testcase is not actively listening, an error is reported. Sample 6-32 shows how an RS-232 monitor can be modified to detect if a serial transmission is started by the design before the testcase is ready to receive it.

Sample 6-32.
Detecting
usage errors in
a RS-232
monitor.

```
process
begin
    loop
        -- Wait for testcase or serial Tx
        wait on to_srv'transaction, Tx;
        exit when not Tx'event;
        assert FALSE
            report "Missed activity on Tx"
            severity ERROR;
    end loop;
    -- Wait for the serial Tx to start
    wait on Tx;
    ...
end process;
```

Response can be
collected for later
retrieval.

Having to continuously call the output monitoring procedures can be cumbersome for the testcase. If the exact timing of output operations is not significant, only their relative order and data carried, the response can be autonomously monitored at all times. The data carried by each operation and an optional description of the operation are queued for later reception by the testcase.

The testcase then is free to introduce delays between the calls to the procedure that retrieves the next response. This procedure returns immediately if a previously received response is available in the queue. It may also block waiting for the next response if the queue is empty. A non-blocking option may be provided, along with a mechanism for reporting that no responses were available. Sample 6-33 shows how the server process for an RS-232 monitor returns responses from a queue implemented using an array. Sample 6-34 shows how the actual monitoring is performed in another process that then puts the received data into the queue. Queues can be implemented using lists, as shown in "Lists" on page 115.

Sample 6-33. Server process in an autonomous RS-232 monitor.

```
process
begin
    wait on to_srv'transaction;
    -- Is queue empty?
    if pop = push then
        wait on push;
    end if;
    frm_srv.data <= queue(pop);
    pop <= (pop + 1) rem queue'length;
end process;
```

Sample 6-34. Monitor process in an autonomous RS-232 monitor.

```
process
begin
    wait until Tx = '1';
    . . .
    -- Is the queue full?
    assert (push+1) rem queue'length /= pop;
    queue(push) <= data;
    push <= (push + 1) rem queue'length;
end process;
```

Autonomous Error Detection

Data may contain a description of the expected response.

In "Packet Processors" on page 215, I described a verification strategy where the data sent through a design carried the information necessary to determine if the response was correct. Autonomous monitors can use this information to detect functional errors. Sample 5-68 shows an example of a self-checking autonomous monitor. The procedural interface of these monitors could provide configuration options to define the signature to look for in the received data stream.

INPUT AND OUTPUT PATHS

Each testcase must provide different stimulus and expect different responses. These differences are created by configuring the test harness in various ways and in providing difference data sequences. This section describes how data can be obtained from external files. It also shows how to properly configure reusable verification components and how to ensure that simulation results are not clobbered by using unique output file names.

Programmable Testbenches

Data was assumed to be hardcoded in each testcase.

Throughout this chapter, there is no mention of the source of data applied as stimulus to the design under verification. Neither is there any mention of the source of expected response for the monitors. In most cases, the stimulus data and the expected response are specified in the verification plan and are hardcoded in the testcase. From the testcase, they are applied or received using bus-functional models in the test harness.

Testbenches can be programmed through external files.

A testcase can be implemented to read data to be applied to, or be expected from, the design under verification from external files. The external files can provide software instructions, a sequence of packets, video images, or sampled data from a previous design. It is a common strategy when the expected response is provided by an external C model, such as is illustrated in Figure 6-11. Programmable testbenches have an advantage: they do not need to be recompiled to execute a new testcase. When compilation or initialization times become critical factors, such as when SDF back-annotation is involved (see "SDF Back-Annotation" on page 305), they offer a technical solution to minimizing the number of time a model is compiled or initialized.

Figure 6-11.
Using external input files for stimulus and response

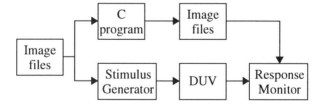

Verilog and VHDL have file-input capabilities.

VHDL is capable of reading any text file, albeit in a very primitive fashion, using the general-purpose routines in the *textio* package. In Verilog, you can only read files of binary or hexadecimal values into a memory, using the *$readmemb* and *$readmemh* system tasks respectively. This requires that the external data representation first be compiled into binary or hexadecimal values before being read into a Verilog testbench. The testbench then interprets the numeric codes back into data or instructions.

Verilog's built-in file input tasks read the entire file into a memory. If the file contains large amount of sequential data, a large memory is required, consuming a significant portion of the available computing resources. Using a PLI function to read data in a sequential fashion is a better strategy. Only the required information is kept in memory during the simulation, improving performance. A link to an implementation of a *scanf*-like PLI task can be found in the *resources* section of:

http://janick.bergeron.com/wtb

Configuration Files

Configuration should be controlled by the testcase.

With the flexibility that packaged bus-functional models offer, they also offer the possibility of making testbenches difficult to understand or manage. Each bus-functional model should be self-contained, controlled through its procedural interface only. The functionality and implications of a testcase can be understood only by examining the top-level testcase control description. All of the non-default configuration settings originate from that single point.

Avoid using external configuration files.

The more files required to make a testcase complete, the more complicated the management task to reproduce a particular configuration. The complexity of file management grows exponentially with the number of files. External files should only be used for large data sequences, either for stimulus or comparison. Short files containing configuration information should be eliminated in favor of specifying the configuration in the testcase itself. Sample 6-35 shows an example of improper configuration using an external file. The random number generator should be seeded by the testcase control, using the task shown in Sample 6-36.

Sample 6-35. Improper configuration using an external file.

```
initial
begin: init_seed
    integer seed[0:0];

    $readmemh("seed.in", seed);
    $random(seed[0]);
end
```

Sample 6-36.
Configuration
using a proce-
dural inter-
face.

```
task seed;
   input [31:0] init;

   $random(init);
endtask
```

Make filenames
configurable.

If you must use a file for reading input stimulus or data to compare against, do not hardcode the name of the file that must be used to provide this information in the bus-functional model. If a hard-coded pathname is used, it is not obvious, from the testcase control, that a file is used. If a filename must be specified through a procedural interface, it is immediately apparent that a file is used to execute a testcase. Sample 6-37 shows how a filename can be specified through a procedural interface in VHDL using the *string* type. The same flexibility can be provided in Verilog by simply allocating 8 bits per characters in a task input argument. Sample 6-38 shows an example and the section titled "Output File Management" on page 309 has more details.

Sample 6-37.
User-speci-
fied filename
in VHDL.

```
procedure data_from_file(name: in string) is
   file fp: text is in name;
begin
   ...
end data_from_file;
```

Sample 6-38.
User-speci-
fied filename
in Verilog

```
task data_from_file;
   input [8*32:1] name;

   $readmemh(name, mem);
endtask
```

Concurrent Simulations

Make sure filena-
mes are unique.

There is another problem with using hardcoded pathnames. If multiple simulations must be run concurrently, a hardcoded filename creates collisions between two simulations. Each simulation tries to produce output to the same file, or read data from the same file. Each simulation must be able to run without conflicting with each other. Therefore, the filenames used for each testcase must be unique.

This problem is typically encountered when generating waveform trace files. By default, the trace information goes to a file with a generic name, such as *verilog.dump* for VCD dump files. To guarantee that each testcase uses a different file, provide a user-specified filename that includes the name of the testcase. Sample 6-39 shows how a string *parameter* containing the testcase name in Verilog can be concatenated to a string literal to create a full filename. In VHDL, the concatenation operator would be used between two *string* expressions, as shown in Sample 6-40.

Sample 6-39.
Generating
unique filenames in Verilog.

```
parameter testcase = "...";

initial
begin
    $dumpfile({testcase, ".dump"});
    $dumpvars;
end
```

Sample 6-40.
Generating
unique filenames in VHDL.

```
architecture test of bench is
    constant testcase: string = "...";
begin
    process
    begin
        read_from_file(testcase & ".dat");
        ...
    end process;
end test;
```

Compile-Time Configuration

Avoid using compile-time configuration of bus-functional models.

When a language offers a preprocessor, it is often used as the mechanism for configuring source code. A different configuration requires a recompilation of the source code using a different *header* file. With most Verilog simulators always recompiling the source code before each simulation, it is a technique that appears efficient and is easy to use. This technique should be discouraged to minimize the compilation requirements in compiled simulators or languages. It also makes managing a testcase more complicated as an additional file, separate from the testcase control, must be managed and kept up-to-date. Furthermore, it may be impossible to ensure the uniqueness of the header file name for each testcase configured by the preprocessor. Using compile-time configuration may make it impossible to run concurrent compiled simulations.

Most compile-time configurations are not modified for a specific testcase.

To minimize the number of files used in a testbench, a single compile-time configuration file is usually used. It contains the definitions for all configurable preprocessor symbols in the test harness. The majority of them have identical values from testcase to testcase, with only a few taking different testcase-specific values. Instead, providing a default value for the configuration parameters that do not need a specific value for a given testcase would avoid the needless duplication of information. Sample 6-41 shows an example of configuring the maximum time-out value of a watchdog timer using an external header file (shown in Sample 6-42) which defines the appropriate preprocessor definitions. Sample 6-43 shows how to provide the same configurability through a procedural interface. A sensible default value is provided that can be used by most testcases, requiring no specific configuration instructions.

Sample 6-41.
Compile-time configuration.

```
module watchdog;
`include "defs.vh"

integer count;
initial count = 0;
always @ (posedge clk)
begin
    count = count + 1;
    if (count > `TIMEOUT) ...
end
endmodule
```

Sample 6-42.
Compile-time configuration definition.

```
`define TIMEOUT     1000000
`define CLK_PERIOD 100
`define MAX_LENGTH 500
`define DUMPFILE    "testcase.dump"
```

Plan your configuration mechanism.

If different configurations must be verified with similar stimulus and response, consider architecting your testcase to facilitate its configuration from within the testbench. The next section should provides some useful techniques.

VERIFYING CONFIGURABLE DESIGNS

This section describes how to verify two kinds of design configurability: soft and hard.

Sample 6-43.
Equivalent
procedural
configuration.

```
module watchdog;

integer count, max;
initial
begin
   count = 0;
   max    = 32'h7FFF_FFFF;
end

task timeout;
   input [31:0] val;

   max = val;
endtask

always @ (posedge clk)
begin
   count = count + 1;
   if (count > max) ...
end
endmodule
```

Soft configuration
can be changed by
a testcase.

There are two kinds of design configurability. The first is *soft* configurability. A soft configuration is performed through a programmable interface and can be changed during the operation of the design. Examples of soft configurations include the offsets for the almost-full and almost-empty flags on a FIFO, the baud rate of a UART, or the routing table in a packet router. Because it can be modified during the normal operation of a design, soft configuration parameters are usually verified by changing them in a testcase. Soft configuration is implicitly covered by the verification process.

Hard configuration
cannot be changed
once simulation
starts.

The second kind of configurability is *hard* configuration. It is so fundamental to the functional nature of the design, that it cannot be modified during normal operations. For example, whether a PCI interface operates at 33 or 66 MHz is a hard configuration. So is the width and depth of a FIFO, or the number of master devices on an on-chip bus. Hard configuration parameters are constant for the duration of the simulation and often affect the testbench as well as the design under verification. A testbench must be properly designed to support hard configuration in a reproduceable fashion.

Configurable Testbenches

Configure the test-
bench to match the
design.

If a design can be made configurable, so can a testbench. The configuration of the testbench must be consistent with the configuration of the design. Using a configuration technique similar to the
one used by the design can help ensure this consistency. Using
generics or *parameters* to configure the testbench and the design
allows the configuration defined at the top-level to be propagated
down the hierarchy, from the testcase control, through the test harness, and into the design under verification.

Parameters and
generics can con-
figure almost any-
thing.

Generics and *parameters* were designed to create configurable
models. They offer the capability to configure almost any declaration in a VHDL or Verilog testbench. You can use them to define
the width of a data value, the length of an array, or the period of the
clock signal. In VHDL, they can even be used to determine the
number of instantiated devices using a *generate* statement. Wherever a constant literal value is used in a testbench, it can be replaced
with a reference to a *generic* or a *parameter*.

Generics and
parameters are part
of a component
interface.

If a testbench component is configurable using generics or parameters, they become part of its interface. Whenever a configurable
component is used, the value of the generics or parameters must be
specified, if they must differ from their default values.

In VHDL, this is accomplished using the *generic map* construct in
an instantiation or configuration statement. In Verilog, it is done
using the *defparam* statement or the *#( )* construct in an instantiation
statement.

Sample 6-44 shows the interface of a memory model with configurable numbers of address and data bits. To use this model in a
board or system under verification, the *AWIDTH* and *DWIDTH*
parameters must be properly configured. Sample 6-45 shows both
methods available in Verilog (I prefer the first one because it is self-
documenting and robust to changes in parameter declarations.)
Sample 6-46 shows how to map generics in VHDL. Notice how the
configurability of the system-level model is propagated to the
memory model.

Sample 6-44.
Configurable
memory
model.

```
module memory(data, addr, rw, cs);
parameter DWIDTH = 1,
         AWIDTH = 1;
    inout [DWIDTH-1:0] data;
    input [AWIDTH-1:0] addr;
    input rw;
    input cs;
...
endmodule
```

Sample 6-45.
Using a con-
figurable
model in Ver-
ilog.

```
module system(...)
parameter ASIZE = 10,
         DSIZE = 16;
wire [DSIZE-1:0] data;
reg  [ASIZE-1:0] addr;
reg              rw, cs0, cs1;

memory m0(data, addr, rw, cs0);
defparam m0.AWIDTH = ASIZE,
         m0.DWIDTH = DSIZE;
memory #(DSIZE, ASIZE) m1(data, addr, rw, cs1);
endmodule
```

Sample 6-46.
Using a con-
figurable
model in
VHDL.

```
entity system is
    generic (ASIZE: natural := 10;
             DSIZE: natural := 16);
    port (...);
end system;

use work.blocks.all;
architecture version1 of system is
  signal data: std_logic_vector(DSIZE-1 downto 0);
  signal addr: std_logic_vector(ASIZE-1 downto 0);
  signal rw, cs: std_logic;
begin
  M0: memory generic map (DWIDTH => DSIZE,
                          AWIDTH => ASIZE)
                port map (data, addr, rw, cs);
  end version1;
```

Top Level Generics and Parameters

Top-level mod-
ules and entities
can have generics
or parameters.

Each configurable testbench component has *generics* and *parameters*, defined by the higher-level module or architecture that instantiates them. Eventually, the top-level of the testbench is reached. The top-level module or entity in a design has no pins or ports, but it can have parameters or generics. The top-level of a testbench can be configured using the same mechanisms as the lower-level components.

For example, Sample 6-47 shows the top-level module declaration for a testbench to verify a FIFO with a configurable width and depth. Notice how the module does not have any pins, but does have parameters.

Sample 6-47.
Configurable
top-level of a
FIFO test-
bench.

```
module fifo_tb;
parameter WIDTH = 1,
          DEPTH = 1;
...
endmodule
```

Top-level generics
or parameters need
to be defined.

By definition, the top-level is not instantiated anywhere. It is the very top level of the simulation. How can its parameters or generics be set? Some simulation tools allow the setting of top-level generics or parameters via the command line. However, a command line cannot be archived. How then can a specific configuration be reproduced later or by someone else? Wrapping the command line into a script is one solution. But it may not be portable to a different simulator that does not offer setting top-level parameters or generics via the command line.

Use a *defparam*
module in Verilog.

In Verilog, the top-level parameters can be set using a *defparam* statement and absolute hierarchical names. A configuration would be a module containing only a *defparam* statement, simulated as an additional top-level module. Sample 6-48 shows how the testbench shown in Sample 6-47 can be configured using a configuration module.

Sample 6-48.
Configuration
module for
FIFO test-
bench.

```
module fifo_tb_config_0;
    defparam fifo_tb.WIDTH = 32,
             fifo_tb.DEPTH = 16;
endmodule
```

Use an additional
level of hierarchy
and configuration
unit in VHDL.

In VHDL, the *configuration* unit does not allow setting top-level generics. To be able to set them, an additional level of hierarchy must be added. As Sample 6-49 shows, it is very simple and not specific to any testbenches since no ports or signals need to be mapped. The configuration unit can then be used to configure the generics of the testbench top-level, as shown in Sample 6-50.

Sample 6-49.
Additional
level of hierar-
chy to set top-
level generics.

```
entity config is
end config;
architecture toplevel of config is
    component testbench
    end component;
begin
    tb: testbench;
end toplevel;
```

Sample 6-50.
Configuration
unit for FIFO
testbench.

```
configuration conf_0 of config is
for toplevel
    for tb use entity work.fifo_tb(a)
        port map(WIDTH = 32,
                 DEPTH = 16);
    end for;
end for;
end conf_0;
```

SUMMARY

This chapter focused on the implementation of testbenches for a device under verification. It described an architecture that promotes reusing verification components. The portion of the testbenches that is common between all testcases is structured into a test harness. Each testcase is then implemented on top of the test harness, using a procedural interface to apply stimulus to and monitor response from the device under verification. Although external data files can be used, the configuration of bus-functional models by each testcases should be limited to using the available procedural interfaces.

SIMULATION MANAGEMENT

Simulation must
be managed.

In "Revision Control" on page 47, I described how tools can help manage the source code generated by the design team. In "Issue Tracking" on page 52, I described how issues and bugs can be tracked to ensure they are resolved. In this chapter, I address the simulation management issues. We see how to efficiently debug your testbenches using behavioral models. Often overlooked, but important topics, such as terminating your simulation, reporting error, and determining success or failure are covered. We also discuss configuration management: how do you know you are simulating what you think you are simulating?

BEHAVIORAL MODELS

This section desmonstrates how behavioral models can benefit a design project. These benefits can only be realized if the model is written with the proper perspective. This section also shows how to properly model exceptions and explains how to demonstrate a behavioral model to be equivalent to an RTL model.

Testbenches
need a model to
be debugged.

You have decided which testcases are needed to functionally verify a design. Your best verification engineers are developing the test harness. Other engineers are working on writing testbenches or specifying utility packages. A couple of basic testcases, using only low-level functionality in the test harness, are already complete. Hardware design engineers are furiously working on the RTL model, but it will not be available for several weeks. Meanwhile, the test harness and testbenches continue to be written. When all will be said and done, the amount of code written for the verification will surpass the amount of RTL code. You are looking at writing thousands of lines of code without being able to debug them.

Behaviorial models are used to debug testbenches.

What if someone walked up to you and offered you a model, available about at the same time as the first testcases, that runs one hundred times faster than the RTL model and that looks and feels just like the real thing? You could start debugging your test harness and testcases even before the RTL is ready. Because this model simulates faster, the debug cycles would be shorter. By the time the RTL is available to simulate, you'd probably have most of your testcases implemented and debugged. The design schedule could be shortened and the verification would no longer be squarely on the critical path. Sound too good to be true? I'm offering exactly such a model: it is called a *behavioral* model.

Behavioral versus Synthesizable Models

Behavioral models are not synthesizable.

Many books, companies, and individuals used the term "behavioral" to describe a synthesizable model. This book uses the term differently. A model that can be automatically translated into a gate-level implementation by a synthesis tool, such as Synopsys' *Design Compiler*, is called *Register-Transfer-Level* or *RTL* model. It may also be called a *synthesizable* model. This book uses the term *behavioral* model to identify models that describe the black-box functionality of a design. The *Virtual Socket Interface Alliance* uses the term *functional* model.

Behavioral code is not just for test-benches.

In "Behavioral versus RTL Thinking" on page 83, I described the characteristics of behavioral code compared with synthesizable code. Using behavioral descriptions for testbenches is easily acceptable by most design engineers. After all, the testbench will never be implemented in hardware so they never give any thought as to how they would go about it. Their mind hasn't been influenced by an implementation architecture or a synthesizable description of the testbench's functionality. They are still open to describing this functionality using behavioral code.

Writing a behav-
ioral model
requires a differ-
ent mindset than
RTL.

Writing a truly behavioral model of a design requires a greater mental leap. You may have already started to think of a design's functionality in terms of state machines, datapaths, operators, memory
interfaces, and other implementation details. This mindset can be
created simply because the functional specification document was
written with these implementation details in mind. To write a
proper behavioral model, you have to focus on the *functionality*, not
the implementation. If the implementation starts to color your
thinking, you'll simply write what I call an "*RTL++*" model.

Example of Behavioral Modeling

"RTL++" models
may be synthesiz-
able using behav-
ioral synthesis.

For example, consider the specification in Sample 7-1. How would
you write a behavioral description of this functionality? Most write
something similar to the description shown in Sample 7-2. This
description is clearly not synthesizable using logic synthesis tools.
However, it happens to be synthesizable using behavioral synthesis
tools such as Synopsys' *Behavioral Compiler*. The design is behaviorally synthesizable because the description was tainted by the
specification: there is an implicit state machine and everything happens at the active edge of the clock.

Sample 7-1.
Specification
of a debounce
circuit

*The debounce circuit samples the input at every clock cycle. The
debounced version of the input changes state only when eight
consecutive samples of the input have the same polarity.*

Sample 7-2.
"RTL++"
description of
debounce cir-
cuit

```
reg debounced;
always @ (posedge clk)
begin: debounce
    if (bouncing != debounced) begin
        repeat (7) begin
            @ (posedge clk);
            if (bouncing == debounced)
                disable debounce;
        end
        debounced <= bouncing;
    end
end
```

A behavioral model cannot be refined into a synthesizable model.

The objective of a behavioral model is to faithfully represent the functionality of a design, in a way that is easy to write and simulate. The behavioral model is designed to help verification, and indirectly, the implementation. When properly written, it cannot be refined into a synthesizable model.

For example, what is the <u>functionality</u> of the debounce circuitry specified in Sample 7-1? It prevents pulses on the primary input, narrower than 8 clock periods, from making it to the debounced output. It is similar to a buffer with a significant inertial delay. This behavior can be modeled using a single statement in both Verilog and VHDL, as shown in Sample 7-3 and Sample 7-4. They use the inertial delay model built in each language. If required, please refer to a suitable Verilog or VHDL reference book[1] for a detailed description of inertial delays.

Sample 7-3. Behavioral description of debounce circuitry in Verilog

```
assign #(8*cycle) debounced = bouncing;
```

Sample 7-4. Behavioral description of debounce circuitry in VHDL

```
debounce: debounced <= bouncing after 8 * cycle;
```

Delays cannot be synthesized.

The descriptions in Sample 7-3 and Sample 7-4 are far from being synthesizable. It is not possible to synthesize a specific inertial delay. The other limitation of these descriptions is the need to know the clock period. It could be specified using a *constant*, a *generic*, or a *parameter*, but the behavioral model would not adjust to different clock periods as the real implementation would. If this is an important requirement, the clock period could be determined at runtime by sampling two consecutive edges. Sample 7-5 shows how this could be performed. Notice how the clock cycle is measured

1. Titles have been suggested in the Preface on page xix.

only once to improve simulation performance. It is unlikely that the clock period will change significantly during a simulation. Computing the clock period at every clock cycle would simply consume simulation resources without accomplishing additional work.

Sample 7-5.
Measuring the
clock period in
the debounce
circuitry

```
architecture beh of debounce is
    signal cycle: time := 8 * 10 ns;
begin
    process
        variable stamp: time;
    begin
        wait until clk = '1';
        stamp := now;
        wait until clk = '1';
        cycle <= now - stamp;
        wait;
    end process;

    debounced <= bouncing after 8 * cycle;
end beh;
```

Characteristics of a Behavioral Model

They are partitioned for maintenance.

A behavioral model is partitioned differently from a synthesizable model. The latter is partitioned to help the synthesis process. Partitioning is decided along implementation lines, producing a design with several instances arranged in a wide and shallow structure.

Behavioral models are partitioned according to the whim of the author. It tends to be partitioned according to main functional boundaries to avoid maintaining one large file, or to allow more than one author to write it. Duplication of function in a model, such as many interfaces of the same type, is also implemented using multiple instances of a single description. Behavioral models tend to have very few instances creating a narrow and shallow structure of large blocks.

They do not use a clock.

A clock signal is an implementation artifice for synchronous design methodologies. They are functionally irrelevant. A behavioral model does not change state synchronously with a clock. Instead, it uses many different synchronization mechanisms. While an RTL model continuously recomputes and updates the value of inferred registers, a behavioral model performs computations only when necessary.

Consider the RTL model in Sample 4-3 on page 86: The process labelled *SEQ* is executed every time the clock changes. The signal named *STATE* is assigned at every rising edge of the clock signal, regardless of the value of *NEXT_STATE*.

The equivalent behavioral model in Sample 4-4 on page 86, on the other hand, does not even use the clock. Instead, it acts on the only functionally significant event: the change in *ACK*. It changes the only functionally significant state, the state of the *REQ* ouput.

A clock would only be used when data needs to be sampled or produced synchronously with a clock signal. Examples of synchronous interfaces include the PCI bus, or the Utopia ATM interface. The clock signals for synchronous interfaces are usually externally generated and are not used any further by the behavioral model.

They do not use FSMs.

Synthesizable models are littered with Finite State Machines. They are the primary synchronous design mechanism for implementing control algorithms. When writing software using a language like C++, you would not consciously implement it as a series of cooperating Finite State Machines. The language does not lend itself very well to that.

Instead, the control algorithm and the data transformations would be part of the control flow of the program. Its state would depend on the current values of the variables and the location of the statement under execution in the program sequence.

Behavioral models follow a similar strategy. Consider the example in Sample 4-3 on page 86. The state of the RTL model is determined by the value of the state register and the current input values. The same code is executed over and over. On the other hand, the state of the behavioral model shown in Sample 4-4 on page 86 depends only on which *wait* statement is currently being executed.

Data can remain at a high-level of abstraction.

The skills of the hardware engineer reside in mapping a complex functionality into the very primitive resources available in hardware. Everything must be implemented using a binary value, with a small number of bits, and reduced to integer arithmetic. A behavioral model can make use of the high-level data types provided by the language, such as floating-point numbers, records, and multidimensional arrays. The section titled "Data Abstraction" on page 100 illustrates many examples of using high-level data abstraction instead of representations suitable for implementation.

Data structures are designed for ease-of-use, not implementation.

In a synthesizable model, the format of the data structures are organized to make implementation possible. For example, imagine a routing table in a packet router is logically composed of 256-bit records with various fields. The router is specified to support 1024 possible routes and the table is maintained by an external processor through a 16-bit wide interface.

The physical implementation of the routing table is likely to use a 16-bit RAM with 16k locations. Whenever the routing engine performs a table look-up, it has to read a block of sixteen words to build the entire 256-bit routing record.

If the table maintenance via the CPU interface has a much lower frequency than packet routing, a behavioral model would instead optimize the data structure for the table look-up and routing operation. The routing table would be implemented using an array of records with 1024 locations. It would also probably use a sparse array implementation to minimize memory usage as well. The table would look the same from the CPU's perspective, with each 16-bit access being performed at the right offset within the record identified by the upper 10 bits of addresses. Sample 7-6 shows a Verilog implementation of the CPU access into the routing table of the behavioral model.

Sample 7-6.
Mapping a
narrow access
in a wide data
structure

```
reg [255:0] table [0:1023];

always
begin: cpu_access
   reg [255:0] entry;
   ...
   entry = table[addr[13:4]];
   if (read) data = entry >> addr[3:0];
   else begin
      for (i = 0; i < 16; i = i + 1) begin
         entry[addr[3:0]*16+i] = data[i];
      end
   end
   ...
end
```

Their interfaces are implemented using bus-functional models.

The testbench is a behavioral model of the environment. To make implementation more efficient, Chapters 5 and 6 explained how bus-functional models are used and located in a testcase-independent test harness. The bus-functional models abstract data from the physical level to a functional level where they are simpler to process using behavioral code.

The same strategy can be used when writing a behavioral model. Bus-functional models are used for each interface around the periphery of the model. Data is transformed behaviorally and moved from bus-functional model to bus-functional model according to the function of the device. And as Figure 7-1 shows, you will likely be able to reuse the bus-functional models written for the testbench in your behavioral model.

Figure 7-1.
Stucture of a
UART test
harness and
behavioral
model

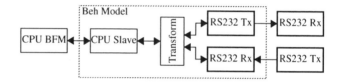

Modeling Reset

Reset is part of the RTL coding style.

Modeling exceptions can take of lot of time and introduce a lot of intricacies in an otherwise simple algorithm description. When writing a synthesizable description, modeling the effect of reset on the state elements is defined in the supported coding style. For example, Sample 7-7 shows how an asychronous reset is modeled to reset a Finite State Machine. Resetting an entire RTL model is accomplished by having each process infer a register included in the logic to handle the reset exception.

Sample 7-7.
Modeling an
asynchronous
reset in RTL

```
process (clk, rst)
begin
   if rst = '1' then
      state <= IDLE;
   elsif clk'event and clk = '1' then
      case state is
         ...
      end case;
   end if;
end process;
```

Behavioral models must reset variables and execution points. As described in the previous section, the state of a behavioral model is not just composed of the values of the variables. It also includes the location of the statement currently being executed in the sequence of statements describing each process. To reset a behavioral model, you need not just reset the content of the variables. You must also reset the execution to a specific statement, usually at the top of the process. For example, resetting the process shown in Sample 7-8 would require resetting the variables and signal drivers to their initial values, as well as restarting the execution of the process at the top.

Sample 7-8.
Behavioral
process have
to be reset

```
process
    variable count: integer := 0;
begin
    strobe <= '0';
    wait until go;
    while (go) loop
        count := count + 1;
        wait on sync;
    end loop;
    strobe <= '1';
    wait for 10 ns;
    strobe <= '0';
    wait until ack = '1';
    count := 0;
end process;
```

In VHDL, check for exceptions in all *wait* statements. *Processes* in VHDL can be affected by other processes only through signals. For a process to be reset, it has to monitor a reset signal, then take the appropriate action once reset is detected. A single reset signal of type *boolean* is sufficient. It would be set to *true* by a reset control *process* whenever a valid reset condition from any number of sources - such has a hardware, power-up or software reset - is detected.

Using a *boolean* type avoids any misunderstanding about the active level of reset. The activity level of a signal, either high or low, is an implementation detail that we need not concern ourselves with internally. Sample 7-9 shows the process from Sample 7-8 with reset detection and handling. Pretty ugly and unmaintainable if you ask me. An otherwise straightforward sequential description gets turned into a complex network of nested *if* statements.

Sample 7-9.
Behavioral
process with
reset detec-
tion and han-
dling.

```
process
   variable count: integer := 0;
begin
   strobe <= '0';
   wait until go or reset;
   if not reset then
      while (go and not reset) loop
         count := count + 1;
         wait until sync'event or reset;
      end loop;
      if not reset then
         strobe <= '1';
         wait until reset for 10 ns;
         if not reset then
            strobe <= '0';
            wait until ack = '1' or reset;
         end if;
      end if;
   end if;
   count := 0;
end process;
```

In VHDL, embed
the process body
in a *loop* state-
ment.

In VHDL, the best way to reduce the clutter of nested control flow statements is to embed the body of the process into an infinite loop, as shown in Sample 7-10. The loop iterates during normal operations but is exited whenever a reset condition is detected. The implicit loop around the *process* statement takes the execution of the process back to the top where the initialization code is located. Sample 7-11 shows the resetable process shown in Sample 7-9 with this new control structure. Each *wait* statement must still detect the reset condition, but the sequential description of the algorithm remains almost untouched.

Sample 7-10.
Structure of a
behavioral
process with
reset detec-
tion and han-
dling

```
process
begin
   -- Initialization
   main: loop
      -- Body of process
      ...
      exit main when reset;
      ...
   end loop main;
end process;
```

Sample 7-11.
A structured
behavioral
process with
reset detec-
tion and han-
dling

```
process
    variable count: integer;
begin
    main: loop
        count := 0;
        strobe <= '0';
        wait until go or reset;
        exit main when reset;
        while (go) loop
            count := count + 1;
            wait until sync'event or reset;
            exit main when reset;
        end loop;
        strobe <= '1';
        wait until reset for 10 ns;
        exit main when reset;
        strobe <= '0';
        wait until ack = '1' or reset;
        exit main when reset;
    end loop main;
end process;
```

In Verilog, disable
all the blocks.

Resetting a behavioral model in Verilog is much more elegant and easy. When an exception is detected, all you need to do it disable all the blocks in the model using the *disable* statement. The *always* blocks restart their execution from the top.

Replace *initial*
blocks with *always*
blocks.

Only the *initial* blocks present a difficulty. Since they only run once in a simulation, they cannot be disabled since they are no longer active. If they are still active, disabling them would simply make them inactive immediately. To include *initial* blocks in the reset handler, simply replace them with *always* blocks with an infinite *wait* statement at the bottom. Sample 7-12 shows an original Verilog behavioral model. Sample 7-13 shows the same model, this time with the proper handling of reset exceptions using the *disable* statement.

Encapsulate the
disable statements
in a task.

It is good practice to encapsulate all *disable* statements into a single task to perform a reset of a Verilog behavioral model. Multiple reset sources and exception detection can call this task to perform the reset operation. It also reduces maintenance to a single location when *always* blocks are added or removed. The *reset* task can also be called using a hierarchical name when a higher-level module in a complex behavioral model needs to reset all its lower-level components. This is more efficient than having to assert a reset signal

Sample 7-12.
Behavioral
model in Ver-
ilog

```
initial count = 0;
always
begin
   strobe <= 1'b0;
   wait (go);
   while (go) begin
      count = count + 1;
      @ sync;
   end
   strobe <= 1'b1;
   #10;
   strobe <= 1'b0;
   wait (ack);
end
```

Sample 7-13.
Behavioral
model with
reset detec-
tion and han-
dling

```
always
begin: init
   count = 0;
   wait (0);
end

always
begin: main
   strobe <= 1'b0;
   wait (go);
   while (go) begin
      count = count + 1;
      @ sync;
   end
   strobe <= 1'b1;
   #10;
   strobe <= 1'b0;
   wait (ack);
end

always
begin
   // Detect reset exception
   ...
   disable init;
   disable main;
end
```

which is broadcasted through the pins of all interfaces in the model.
Sample 7-14 shows the reset handler of Sample 7-13 modified to
use a task to disable all of the blocks.

Sample 7-14.
Encapsulating
the *disable*
statements in a
task

```
task reset;
begin
    disable init;
    disable main;
end
endtask

always
begin
    // Detect reset exception

    ...
    reset;
end
```

Writing Good Behavioral Models

Many attempts to
write behavioral
models fail.

I have seen and heard of many projects where the use of behavioral models was attempted, but without producing much benefit over RTL models. Often, the behavioral model was abandoned in favor of the RTL model as soon as the latter became available. The behavioral model failed to exhibit any the benefits outlined in "The Benefits of Behavioral Models" on page 286.

Writing a good
behavioral model
requires special-
ized skills.

Further investigation into those failed attempts usually reveals that the behavioral model was written by experienced hardware design-ers. Unfortunately, their valuable skills were not appropriate to writing good behavioral models. Their level of thinking was still too close to the implementation and they had difficulty thinking in terms of higher levels of abstraction. Very often, there was the implicit intent of refining the behavioral model into a synthesizable model. This is a fatal mistake as it is conducive to low-level think-ing, yielding not a behavioral model, but an "RTL++" model.

Focus on the rele-
vant functional
details.

All the techniques illustrated in this chapter, as well as in Chapter 4, can be used and still yield a poor behavioral model. A good behav-ioral model focuses on the details that are functionally relevant and not on implementation artifices. For example, the *latency* of a design - the number of clock cycles necessary for an input to be transformed into an output - is usually not functionally relevant. If you insist on writing a model that is clock-cycle accurate with the actual implementation, you may be spending a lot of effort and add-ing a lot of complexity for a characteristic that is not functionally relevant.

At first glance, latency seems a significant characteristic.

To many, saying that latency may not be a relevant functional detail and should not be modeled sounds like a recipe for disaster. But if you take a step back from your design, ignoring its implementation details, does it *really* matter whether a particular output comes exactly N cycles after the corresponding input was sampled? As long as the *order* of these outputs is the same, is the time at which they come out significant?

Consider the speech synthesizer design illustrated in Figure 3-4 on page 78. To produce audible speech, coefficients must be modified at regular intervals to produce the different sequences of sounds that compose normal speech.

For example, to say "cat", the coefficients would be modified to create the sequence of sounds "k", "a", "a", "a", "t", "t". From these coefficients, a digitized sound waveform should come out at a 8kHz sample rate. The delay between the time the coefficients are set and the corresponding sound is synthesized is irrelevant, as long as it is under the limit of perception by the user. A similar argument can be made for packet routers: it does not really matter how long it takes for packets to transit through a routing node, as long as they eventually come out in the same order.

In some cases, latency is significant.

The only time where a detail like latency is significant is when the design under verification does not have complete visibility over a system-level "unit of work". A unit of work is the smallest amount of data than can be processed by the system: an atomic operation. For example, a packet router's unit of work is an entire packet. In a speech synthesizer, it is a vocal sound. In a hardware tester, it is a complete vector with input and expected output values. If the design under verification only processes a portion of the unit of work, it is important that the latencies in the reconvergent paths are identical so the unit of work gets properly reassembled.

For example, the input formatter in a hardware tester, as illustrated in Figure 7-2, only processes the input value. For the corresponding expected output value to be checked at the proper time, it must have the exact same latency as the *Expect Delay* design.[2] In a packet router, as illustrated in Figure 7-3, if the packet is dismembered to

---

2. Actually, since the latter is easier to design, its latency is made to match that of the input formatter, whatever that it may be.

be routed by different switching node, each node must have an identical latency for the packet to be properly put back together. If you mix behavioral and RTL models in a system-level verification, and each has a different latency, the system-level simulation would become a very effective packet scrambler!

Figure 7-2.
Reconvergent paths in a hardware tester

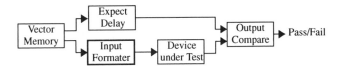

Figure 7-3.
Reconvergent paths in a packet router

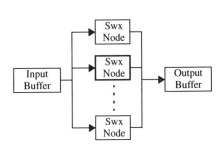

Do not let the testbench dictate what is functionally relevant.

The reason most often cited for making a behavioral model clock-cycle accurate with the implementation is to be able to pass the same cycle-oriented testbenches. If the testbenches enforce a specific latency, they are verifying a specific implementation, not a specification.[3] I hope I have successfully explained how to write testbenches that are independent of the latency of the design under verification in Chapters 5 and 6. If your testbenches do not expect a specific latency, then you need not model it.

Details relevant at the system-level can be back-annotated.

An implementation detail, such as latency, may not be relevant to the functionality of the *stand-alone* design under verification. However, it may be critical for the proper operation of the system-level design. If that is the case, such as the example designs shown in Figure 7-2 and Figure 7-3, the behavioral model may still be modeled as if the latency was not important and perform its transforma-

---

3. Unless of course a specific latency is required, in which case it should be specified in the specification document. And if something is specified, it should be modeled and verified.

tion in zero-time. At appropriate points in the input or output paths, programmable delay pipelines can be introduced so the exact latency of the implementation can be *back-annotated* into the behavioral model. The behavioral model would then model the functionality of the synthesizable model at a clock-accurate level. Sample 7-15 shows a configurable delay pipeline to adjust the latency of a behavioral model.

Sample 7-15.
Configurable
delay pipeline

```
process (clk)
    constant delay: natural := 1;
    type pipeline_typ is array (integer range <>)
        of data_typ;
    variable pipeline: pipeline_typ(1 to delay);
begin
    if clk = '1' then
        actual_ouput <= pipeline(delay);
        pipeline := output &
                        pipeline(1 to delay - 1);
    end if;
end process;
```

Specify the functionality, not the implementation.

Another big obstacle to writing good and efficient behavioral models is the level of the specification for the design. If it is written at a very low level, it becomes difficult to abstract significant functionality and discard irrelevant implementation details. I once had to write a behavioral model for a customer whose functional specification was done using technology-independent schematics using a general-purpose drawing tool. Each block was specified independently with no description of the overall functionality. Not only did it make the job of writing RTL code that met timing requirements difficult, it made writing a high-level behavioral model impossible. After 10 weeks, I had a model that was barely faster than the RTL model. But after those 10 weeks, I was able to piece the entire design together in my mind and understand the intended functionality. I scrapped the first model and rewrote it entirely in under two weeks. That newer model outperformed the RTL model. Had the specification been written at an appropriate level in the first place, a more effective behavioral model could have been written from the start.

Behavioral Models Are Faster

They are faster to write.

As shown in "Behavioral versus RTL Thinking" on page 83, a behavioral model is much faster to write simply because the functionality is described using significantly fewer statements than an RTL model. Furthermore, they do not need to meet physical timing or other implementation constraints. They are written with the sole purpose to describe the functionality of a design.

They are faster to debug.

The fewer statements, the fewer bugs. Bugs are easier to identify because of the simpler descriptions. The code is written based on a functional description. It is not cluttered by directives aimed at a synthesis tool, or twisted to be synthesized into specific hardware structures. They also tend to use fewer parallel constructs, instead preferring large sequential descriptions in a few processes. Sequential code is much easier to debug than parallel code, since it does not involve synchronization or data exchange intricacies.

They are faster to simulate.

Less code used to describe a function should naturally simulate faster. But the greatest contributor to the increase in simulation speed of a behavioral model over a synthesizable model is the synthesizable subset itself. Look at all the *processes* and *always* blocks used to infer registers. Each and every one of them is sensitive to the clock. If you remember the discussion on event-driven simulation in "The HDL Parallel Engine" on page 125, you know that this causes all of these processes to be scheduled for execution after each event on the clock signal, whether their state changes or not.

In a typical ASIC, activity levels are below 40 percent. This means that over 60 percent of the processes are evaluated for no reason. A behavioral model only executes when there is useful work to be done. The load it puts on the simulator is much lower. In the small example illustrated in "Contrasting the Approaches" on page 85, the activity in the behavioral model is estimated to be 20 times lower than in the equivalent RTL model.

They are faster to bring to "market".

Being faster to write and debug, a behavioral model takes significantly less time to develop to a level where it can be used in a system-level model. With behavioral models, you are able to start system-level simulations sooner. Because they also simulate faster, you are able to run more of them, on less expensive hardware.

The Cost of Behavioral Models

Behavioral models require additional authoring effort.

Someone has to write these behavioral models. If you use your existing resources, it means that the coding of the RTL model will be delayed. If you do not want to affect the schedule of the synthesizable model, you will have to hire additional resources to write the behavioral model. Being a completely separate model, it is a task that is easy to parallelize with the synthesis effort. And writing a behavioral model is not as costly as writing an RTL model. A behavioral model, sufficient to start simulating and debugging the testbenches, should not take more than two person-weeks to produce. A complete model with all of the functionality of the design under verification should not take more than five percent of the effort required to write an equivalent RTL model.

The maintenance requires additional efforts.

When was the last time you were involved in a design project where the functional specification did not change? Whenever a functional or architectural change is made, the behavioral model needs to be modified. Often, these modifications are dictated by the RTL model because the technology cannot implement the original design and still meet timing requirements. Some of these implementation-driven changes can be planned for and made easy to modify, such as the latency. More significant changes may require rewriting a significant portion of the behavioral model. Toward the end of a project, when schedule pressure is at its greatest, it often leads to the decision of abandoning the behavioral model in favor of focusing on the RTL.[4] However, most of the modifications to an RTL model are made to meet timing goals and do not affect the functionality of the design, and thus should not require modification of the behavioral model.

The Benefits of Behavioral Models

Audit of the specification.

Most specification reviews I have attended focus on high-level functions and on the spelling and grammatical errors in the document. The missing functional details were often left to be discovered during RTL coding. Decisions regarding these functional details were then usually made according to the ease of implemen-

---

4. An error in my opinion. See the next section titled "The Benefits of Behavioral Models".

tation. There is nothing like writing a model to make you thoroughly read a specification document.

For example, after you've coded a particular function that occurs under some condition, you've come to the *else* part of the *if* statement. What should be done when the condition does <u>not</u> occur? Flip, flip, flip through the specification document. Not a word. You've just found a case of incomplete specification! Since you are writing the behavioral model faster than the RTL model, you'll reach that section of the specification earlier than the RTL designers. By the time the RTL model incorporates this functionality, it will have been specified. A similar process occurs with inconsistencies in the specification. When the RTL is written, there are fewer problems in the specification, and thus takes less time to write.

Development and debug of testbenches in parallel with the RTL coding.

Testbenches are implemented using code, just as RTL models are. If the RTL model requires debugging, so do the testbenches. And since the testbenches now account for over 60 percent of the code volume, they require more debugging than the RTL. Since a behavioral model is available much earlier than the RTL code, you are able to debug the testbenches earlier as well. You are effectively debugging the behavioral model and the testbenches while the RTL is being written. And because the behavioral model simulates faster than the RTL model, the testbenches take less time to debug.

Once the RTL is completed, you will have a whole series of debugged testbenches. Whenever an error is detected, it will likely be due to of an error in the RTL model. If you decide to abandon the maintenance of the behavioral model after the RTL is available, debugging the testbenches (which will also need to be modified whenever the RTL is significantly modified) will take much longer. It is important to maintain the behavioral model to keep reaping its benefits for the entire duration of the project.

System verification can start earlier.

Figure 7-4 shows a design process that uses behavioral models for developing the testbenches and the functional verification of the system. Figure 7-5 shows a comparative timeline for a design and verification process with and without behavioral models. The design process is somewhat shortened by using a behavioral model because the testbenches are already debugged. But the greatest saving comes from system verification. The behavioral model is available sooner than the RTL model, so functional verification can start much earlier. Because a behavioral model is much smaller and simulates more efficiently than the equivalent RTL model, you are able to create models of larger systems, execute longer testcases, and run on ordinary hardware platform configurations. If the behavioral model is demonstrated to be equivalent to the RTL model, the latter never needs to be brought into the system-level verification. For systems incorporating very large ASICs, a behavioral model may be that which makes system verification even possible.

Figure 7-4.
Design process including behavioral model

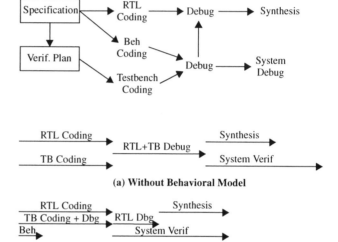

Figure 7-5.
The effect of behavioral models on a project timeline

It can be used as an evaluation and integration tool by your customers.

If your design is to be available as reusable intellectual property or a chipset, a behavioral model can be a powerful marketing tool. Since it only describes functionality, not implementation, and is far from being synthesizable, it should not convey intellectual property information.[5] A customer could start using the behavioral model while the legal issues with licensing the RTL model are being resolved. The system-level models could be used as application notes. The behavioral model could be used to start the integration of your design into your customer's design. Since reusing intellectual property is about time-to-market, a behavioral model can be an effective tool to help your customers improve the odds that they will reach their market window.

---

5. Unless the intellectual property is in the function itself, such as a DSP algorithm.

Demonstrating Equivalence

The RTL and behavioral models must be equivalent.

The greatest benefit from creating a behavioral model comes from system verification. To use it instead of the RTL model in a simulation or as a marketing tool, you have to demonstrate that both are an equivalent representation of the design. I use the term "*demonstrate*" because I do not think it will ever be possible to mathematically *prove* that they are equivalent.

Equivalence checking can prove that a RTL model is equivalent to a gate-level model or to another RTL model because they are structurally very similar. A properly-written behavioral model would use a completely different modeling approach that would be very difficult to mathematically correlate with the equivalent RTL model.

Demonstrate equivalence by using the same test suite.

The only way to demonstrate that the behavioral and the RTL models are equivalent is to verify both of them using the same test suite. If both models pass the same testcases, from a system-level perspective, it should not matter which one you are using. For a testcase to be executable on both models, it must not depend on a specific implementation. Based on the testcase taxonomy described in "Functional Verification Approaches" on page 11, only black- and grey-box testcases can be used to demonstrate equivalence. Both are executed through the same physical interface. Both do not depend on a particular implementation of the design under verification. The grey-box testcases may not be very relevant to the behavioral model as they are designed to test a particular implemenation-specific feature in the RTL model, but should nonetheless execute succesfully.

PASS OR FAIL?

This section describes how the ultimate failure or success of a self-checking testbench is determined.

The absence of error is not a sufficient condition.

The goal of a testbench is to determine if the design under verification passes or fails a testcase. But how do you determine if the design passed the testcase? Is it by the absence of error messages? What if the testcase never ran at all? It could be caused by a lack of licenses, or a run-time error such as running out of memory or experiencing a power failure, or a simple syntax error in your source code. You need positive proof that the testcase successfuly ran to completion.

Produce and look for a termination message.

Do not rely on a time bomb to terminate your testcase. Nor should you attempt to have the simulation terminate by itself through event starvation. Each testcase should be intentionally terminated. Upon termination, it should produce a message that the simulation was terminated normally. If that message is not present, you must assume that the testcase did not run to completion and failed. To terminate a simulation from within the testbench, use the *$finish* statement in Verilog, or the *assert* statement in VHDL. Sample 7-16 and Sample 7-17 illustrate their respective use.

Sample 7-16.
Terminating a Verilog simulation

```
initial
begin: test_procedure
    . . .
    $write("Simulation terminated normally\n");
    $finish;
end
```

Sample 7-17.
Terminating a VHDL simulation

```
test_procedure: process
begin
    . . .
    assert false
        report "Simulation terminated normally"
        severity failure;
end process test_procedure;
```

An error in the testbench could prevent error detection.

What if there is a functional problem in your testbench? That error could prevent the testbench from detecting any errors at all. This would clearly be a *false-positive* situation. You should always ensure that your testbench is functionally correct as part of your testcases. Error detection can be verified by deliberately injecting errors in the design under verification. These errors can be introduced by simply misconfiguring the design for the expected output. For example, a UART could be configured with the wrong parity setting to verify that the output monitor detects the bad parity.

Use errors as a valid response.

Once you decide to inject an error into your design, how should you handle it? You might want to configure your monitor to expect that error and produce a message if it <u>does not</u> see the erroneous condition. I consider this option risky. It increases the complexity of the output monitor, as it may be very difficult to time the detection of the error accurately. But, more importantly, it still does not provide positive proof that your monitor detects the error. Instead, let the error message be produced during simulation. To distinguish this intentional - and <u>mandatory</u> - error, generate a message that an error condition is expected. For error messages that are expected, produce a different label to easily distinguish them from real errors.

Bracket regions where error messages are expected. Having error messages be part of the success criteria also involves risks. What if the error message you expected to receive was not produced and another unexpected one was issued instead? Unless you were intimately familiar with the workings of the testcase and read all error messages, you would not be able to differentiate this functional failure from a successful testcase execution. To prevent the misinterpretation of errors as expected ones, bracket the regions where errors are expected with messages stating the expectations. Sample 7-18 shows a portion of a simulation output. Notice how the region where two error messages were expected is bracketed by expectation messages. This region should be as narrow as possible. Notice also how the label for the error messages is different if the error is expected or not.

Sample 7-18.
Bracketing
exected errors
in a simulation

```
. . .
EXPECT : 2 errors
  NOTE   : . . .
(error): . . .
(error): . . .
EXPECT : 0 errors
. . .
  ERROR : . . .
. . .
```

Provide consistent error message formats. The success or failure of a testcase would be determined by the presence of the specified number of error messages within the expected regions. The final pass or fail judgment could be made by a script parsing the simulation output log file, counting the unexpected or missing errors. To facilitate the implementation of such a script, use a consistent error format. This is best accomplished by using a message log package that produces consistent headers, as shown in Sample 7-19.

Keep track of success or failure in the log package. By using a single message log package as shown in Sample 7-20, it is possible for the simulation to keep track of its own success or failure by comparing the number of errors encountered against the expectation. By including a simulation termination function, the

<table>
<tr><td>

Sample 7-19.
Simulation log
package

</td><td>

```
module log;

integer expected;
task expect;
    input [31:0] n;
    $write("EXPECT: %0d errors\n", n);
    expected = n;
endtask

task error;
    if (expected > 0) begin
        $write("(error): ");
        expected = expected - 1;
    end
    else $write(" ERROR : ");
endtask

endmodule
```

</td></tr>
</table>

final pass or fail indication can be determined by the simulation, without using a script to parse the output log. Such a package is easy to implement in Verilog, but requires shared variables in VHDL, a VHDL-93 feature.

Using a script to parse the simulation output log is still a good idea.

Using a message log package is not sufficient to determine if a testcase is successful. Other errors could have been generated before the simulation started. The log package can only count error messages during simulation, not before or after. You still need to confirm the presence of the termination message to verify that the testcase was properly executed in its entirety. The output log parsing script can also detect the presence of errors or warnings issued by the simulation management tools, linting tools, syntax errors, elaboration warnings, and other possible error conditions not visible to the simulation. You can find a link to a configurable output log parser script in the *resources* section of:

http://janick.bergeron.com/wtb.

MANAGING SIMULATIONS

Are you simulating the right model?

You've defined your verification task through a verification plan. You have a test harness with many bus-functional models and utility packages. Several testcases using that test harness have been written and you can choose between the RTL and behavioral model

Sample 7-20.
Determining
pass or fail in
the simulation
log package

```
module log;

reg fail;
initial fail = 0;
integer expected;

task expect;
   input [31:0] n;
begin
   if (expected) begin
      $write("*FAIL*: %0d missed errors\n",
             expect);
      fail = 1;
   end
   $write("EXPECT: %0d errors\n", n);
   expected = n;
end
endtask

task error;
end
   if (expected > 0) begin
      $write("(error): ");
      expected = expected - 1;
   end
   else begin
      $write(" ERROR : ");
      fail = 1;
   end
end
endtask

task terminate;
begin
   $write("Simulation %0sED\n",
          (fail) ? "FAIL" : "PASS");
   $finish;
end
endtask

endmodule
```

to simulate them. How do you bring all of these components together in a single simulation? How can you reproduce a simulation? And more importantly, how do you make sure that what you simulate is what will be built?

Configuration Management

A configuration is the set of models used in a simulation.

Configuration management is different from source management. Source management, as described in "Revision Control" on page 47, deals with changes to source files and the set of source files making up a particular release. Configuration management deals with the particular set of models you decide to use in a particular simulation. For a specific design, a single configuration would be composed of a specific testcase, the test harness used by that testcase, and the model of the design to be exercised by the testbench. In a system-level simulation, the configuration would also include that particular mix of models used to populate the system model.

It must be easy to specify a particular configuration.

The only information required to define a particular configuration is the identity of the testcase, the test harness, and the model of the design under verification. The problem is that each configuration component is potentially composed of several source files and design units. Many individuals contribute to the creation of these source files and design units. Their number and names may change throughout the project. It is not realistic to expect every engineer who needs to run a simulation to know exactly what makes up a particular component of the desired simulation. Just as bus-functional models abstract the data from the physical implementation level, configuration management abstracts the details of the structure of a model and the files that describe it.

Use a script to create a configuration.

The most efficient way to abstract the configuration details from the user is to provide a script that expands a testcase name and an abstraction level for the design under verification into their respective simulation components. It is very likely that different scripts have to be used for different design. To simplify the user interface and minimize the amount of repeated information, they infer pathnames and expect particular setup files specific to each design.

For example, Sample 7-21 shows the command line of a script named *sim_design* used to simulate a configuration composed of the testcase named *"basic_tx"* on the behavioral model. It is followed by a configuration composed of the testcase named *"overflow_rx"* on the RTL model.

Sample 7-21.
Configuration
script com-
mand line

```
% sim_design -b basic_tx
% sim_design -r overflow_rx
```

Verilog Configuration Management

There are many
ways of specify-
ing files.

There are five different ways to include a source file into a Verilog simulation:

1. Specify the filename on the command line.

2. Specify the name of a file containing a list of filenames, using the *-f* option.

3. Specify a directory to search for files likely to contain the definition of a missing module, using the *-y* option. The files used in the simulation depend on the +*libext* command-line option.

4. Specify the name of a file that may contain the definition of missing modules, using the *-v* option.

5. Include a source file inside another using the `include directive. The actual file included in the simulation depends on the +*incdir* command-line option.

Use the *-f* option.

Of all the mechanisms for specifying input source files in Verilog, only the *-f* option can be source-controlled and reliably reproduced. Depending on command-line options and search orders is not reliable and makes reproducing a configuration difficult. It may also breakdown if conflicts arise when a model is integrated in another environment where directory, file and module names conflict. The file-of-filenames, or *manifest* file, becomes an integral part of the source code. It is maintained by the author as files are renamed, added, or removed. It removes the user from knowing exactly which files make up a particular model, harness or testcase. All the user needs to know is the name and location of the desired manifest file. Verilog does the rest.

Specify all
required com-
mand-line options.

A manifest file is not limited to containing filenames. It can contain any command-line option. A manifest file should include all command-line options required by a model. For example, gate-level models use the *-y* or *-v* option to locate and load the required modules for the gate models. Specifying all required files, including the model for all the gates used would be impractical. A gate-level model can use several dozen different gates. The set of gates can also differ from synthesis run to synthesis run.

Sample 7-22 shows a manifest file for a gate-level model composed of three files. The models for the gates are located in a directory called "*../gates/lib/tech*". The gate-level models require that the "*FAMILY_08*" symbol be defined.

Sample 7-22.
Command-line
options in a
manifest file

```
top.v
fifo.v
ram.v
+define+FAMILY_08
-y ../gates/lib/tech
+libext+.v
```

Manifest files can
be hierarchical.

Since manifest files can contain any command-line option, they can contain *-f* options. If a model is composed of independently configured units, such as a system model, the lower-level units can be included by including their manifest files in the manifest file of the higher-level design. Similarly, a testcase manifest file would include the manifest file for the required test harness in its own manifest file, using the *-f* option.

For example, Sample 7-23 shows the manifest file for the testcase illustrated in Figure 6-5 on page 234, while Sample 7-24 shows the manifest file for the harness.

Sample 7-23.
Hierarchical
manifest file
for testcase

```
testcase.v
packet.v
+define+BEH_PLL
-f harness.mft
```

Sample 7-24.
Manifest file
for test har-
ness

```
harness.v
i386sx.v
```

Select the design
under verification
using its manifest
file.

In Sample 7-24, the harness did not refer to or include the manifest file for the design under verification. This would have limited the execution of the testcase to a single configuration of the design under verification. Instead, the configuration to be verified by the selected testcase and test harness is specified separately. During the elaboration phase of the Verilog model, the design is associated with the instantiation in the harness, as long as the module and pin names match. Sample 7-25 shows how the testcase would be executed on the behavioral model of a design, then on the RTL model. Each model is assumed to be located in a different directory.

Sample 7-25.
Specifying a
testcase and
design config-
uration

```
% verilog -f tests/testcase.mft -f beh/design.mft
% verilog -f tests/testcase.mft -f rtl/design.mft
```

Use relative path-
names.

Manifest files should be written to work under different locations in a file system. Often, different computers on the same network see the same files under a different mount point. If absolute pathnames are specified, every computer has to have the file system mounted at the exact same location. It may not always be possible, especially if a model is to be used by another business entity with a different computer administration policy. The manifest file shown in Sample 7-26 will not be portable to a different file system. The one shown in Sample 7-27, which uses relative pathnames, will be useable regardless of the mount point of the file system.

Sample 7-26.
Absolute path-
names in man-
ifest files

```
/net/raid/proj/titan/tests/testcase.v
/net/raid/proj/titan/utils/packet.v
+define+BEH_PLL
-f /net/raid/proj/titan/harness/harness.mft
```

Sample 7-27.
Relative path-
names in man-
ifest files

```
tests/testcase.v
utils/packet.v
+define+BEH_PLL
-f harness/harness.mft
```

Relative path-
names may need
correction.

The advantage of absolute pathnames is that they do not imply or require a specific working directory. Relative pathnames in manifest files are interpreted as-is by Verilog. The actual location of the file depends on the current working directory where the simulation is invoked. For example, Sample 7-28 shows the absolute pathnames of the files loaded by Verilog using the manifest file shown in Sample 7-27 if the working directory is "*/home/joe/titan*". Compare with Sample 7-29, which shows the assumed pathnames if the working directory was set to "*/home/joe*". Only one of them produces the desired simulation.

Sample 7-28.
Assumed pathnames from a working directory

```
% verilog -f tests/testcase.mft
/home/joe/titan/tests/testcase.v
/home/joe/titan/utils/packet.v
+define+BEH_PLL
-f /home/joe/titan/harness/harness.mft
```

Sample 7-29.
Assumed pathnames from a different working directory

```
% verilog -f titan/tests/testcase.mft
/home/joe/tests/testcase.v
/home/joe/utils/packet.v
+define+BEH_PLL
-f /home/joe/harness/harness.mft
```

The *-F* option is a good start toward a solution.

The *-F* option, similar to the *-f* option, is not supported by all Verilog tools. But it provides a useful mechanism of overcoming the dependence of relative filenames on the current working directory. If a manifest file is specified using the *-F* option instead of *-f*, the relative pathnames in the manifest file are *prepended* with the path to the manifest file.

For example, the manifest file for a behavioral model shown in Sample 7-30 is used to configure a simulation from two different working directories in Sample 7-31. In both cases, the correct files are assumed by Verilog. However, the *-F* option does not work well when the relative pathnames contain directories or command-line options that include relative pathnames in their arguments.

Sample 7-32 shows how the relative pathnames in a manifest file are interpreted when specifying directories. The *-F* option was a good idea, but its implementation leaves a lot to be desired. Furthermore, at the time this book was written, the *-F* option did not work hierarchically.

Sample 7-30.
Manifest file with relative pathnames

```
design.v
fifo.v
ram.v
```

Writing Testbenches: Functional Verification of HDL Models

Sample 7-31.
Assumed path-
names using
the -*F* option

```
% verilog -F titan/beh/design.mft

titan/beh/design.v
titan/beh/fifo.v
titan/beh/ram.v

% cd titan
% verilog -F beh/design.mft

beh/design.v
beh/fifo.v
beh/ram.v
```

Sample 7-32.
Assumed path-
names using
the -*F* option
with hierarchi-
cal pathnames

```
% verilog -F titan/tests/testcase.mft

/home/joe/titan/tests/tests/testcase.v
/home/joe/titan/tests/utils/packet.v
+define+BEH_PLL
-f /home/joe/harness/harness.mft
```

Use a script to
parse the manifest
files.

The -*F* option is a good idea and its implementation can be fixed by using a script to parse the manifest file and correct any relative pathnames. Any relative pathname would be interpreted as relative to the location of the manifest file, not the current working directory. That way, manifest files can be used, unmodified, from any location in the file system.

Sample 7-33 shows how the manifest file in Sample 7-34 would be interpreted using the manifest parsing script.

You will find an PERL implementation of the manifest parse script in the *resources* section of:

```
http://janick.bergeron.com/wtb
```

Any relative pathname is corrected: filenames, as well as arguments to the -*f*, -*F*, -*y*, -*v*, and +*incdir* options. The script can also allow pathnames containing environment variables. Identifying writeable files would be a useful feature of the manifest parsing script. It would indicate files that are potentially different from the version in the source control system.

Filter out files
included multiple
time.

Since manifest files are written independently from each other, they could include the same source file twice. For example, two different

<table>
<tr><td>

Sample 7-33.
Assumed path-
names using a
manifest pars-
ing script

</td><td>

```
% mverilog -F titan/tests/testcase.mft
titan/tests/testcase.v
titan/tests/../utils/packet.v
+define+BEH_PLL
-f titan/tests/../harness/harness.mft

% cd titan/utils
% mverilog -F ../tests/testcase.mft

../tests/testcase.v
../tests/../utils/packet.v
+define+BEH_PLL
-F ../tests/../harness/harness.mft
```

</td></tr>
</table>

Sample 7-34.
Manifest file
with names
relative to the
manifest file

```
testcase.v
../utils/packet.v
-F ../harness/harness.mft
```

units in a design, each with their own manifest file, could include the source files for a shared component.

Sample 7-35 shows two such manifest files. Both include the file named "*fifo.v*" in the *shared* directory. The manifest parse script should filter out multiple references to the same file. Otherwise, Verilog will abort with an error because of a module being defined more than once.

Sample 7-35.
Multiple file
reference in
different mani-
fest files

master/master.mft:

```
master.v
../shared/fifo.v
```

slave/slave.mft:

```
slave.v
../shared/fifo.v
```

Manifest files need
to work differ-
ently with some
compiled simula-
tors.

Using manifest files for Verilog configuration works properly when using a simulator that compiles the source files before every simulation. It also works with compiled simulators with incremental compilation, such as *VCS* with the *-R* option, if they look at all the source files to determine if any source file needs recompilation.

Others, such as *ModelSim* or the *simv* binary produced by *VCS*, have separate compilation and simulation phases.

The manifest files deal with compile-time as well as run-time options. If the simulation and compilation phases are separated, the manifest files need to be split into compilation and simulation manifest files. The split between run-time and compile-time options may be different for each compiled simulator and could be automatically handled by the manifest front-end script.

Compiled simulations are disconnected from the source files. Using compiled simulation creates a disconnect between the source file and the simulation of the compiled files. What if the source file has changed? How do you make sure that what you are simulating is the proper version of the source files? Configuration management techniques for compiled simulations are outlined in the next section. As all VHDL simulators are compiled simulators, the management of VHDL models must deal with this problem. To eliminate this difficulty and provide a familar interface to experienced *Verilog-XL* users, all compiled Verilog simulators provide an interpreted-like mode where all the source files are recompiled every time and simulated immediately.

VHDL Configuration Management

You may not be simulating what you thought was compiled. All VHDL simulators are *compiled* simulators. During compilation, the individual source files are compiled into libraries and translated to object code. During *elaboration*, a top-level unit is selected and, using the configuration information, a hierarchical model is built by recursively connecting entities and architectures into component instances. The elaborated model is then simulated. A separate command (sometimes two) is used to trigger the compilation and the elaboration. This creates a potential disconnect between the source files and what is ultimately simulated. How do you know that the source files located in your directory are the ones you are simulating?

Use makefiles. The most effective way to assure a compiled simulation is up-to-date is to use makefiles. Makefiles, and the associated *make* program, were created in the mid-seventies to maintain software programs and make sure that the compiled code was always up-to-date.

Makefiles contain dependency rules describing relationships between files. If a file is found to be older than a file it depends on, it is brought up to date using a user-defined command. Nowadays, it is not necessary to know the frustrating syntax[6] of Makefiles. All VHDL and compiled Verilog toolsets can generate a Makefile from a compiled model. Third-party utilities, such as *VMK*, can also automatically generate Makefiles from VHDL or Verilog source code.

Use *make* to invoke the simula-tion.

To ensure that a simulation is always up-to-date, do not invoke the simulator manually. Some source files might have changed and you would be simulating an out-of-date model. Instead, use the *make* command to invoke the simulator. The program ensures that any source files that have changed since the last compilation are recom-piled, in the correct order. There should be a *target* for each testcase available for simulation. The name of the target depends on the tool used to generate the Makefile. Sample 7-36 shows how to invoke *make* using a specific target.

Sample 7-36.
Invoking *make*

```
% make basic_tx
% make overflow_rx
```

The model should report the name and version of files.

For additional confidence and a positive confirmation of the files and version of the files simulated in a compiled model, you should have each architecture report the name and revision number of the file that contained it during compilation. Sample 7-37 shows how to use a concurrent *assert* statement and RCS keywords to perform this task. All of the *assert* statements are executed at the beginning of the simulation, displaying the filename and revision information. Because they are not sensitive to any signals, they do not execute again. Unfortunately, it is not possible to have *packages* report their source file name and version numbers. A compiled Verilog model could use a *$write* statement and an *initial* block to accomplish the same thing, as shown in Sample 7-38.

Use configuration units to define the possible configu-rations of the design.

VHDL supports the concept of model configuration using configu-ration units. Configuration declarations are like assembly instruc-tions for a simulation model. They specify which entity and which architecture of that entity should be used for each component

---

6. I personally would like to have a little chat with whomever picked the TAB character as a significant control character!

Sample 7-37.
VHDL model
reporting its
filename and
revision

```
architecture beh of design is
    ...
begin
    assert false
        report "Configuration: $Header$"
        severity note;
    ...
end beh;
```

Sample 7-38.
Verilog model
reporting its
filename and
revision

```
module design(...);
    ...
initial $write("Configuration: $Header$\n");
    ...
endmodule
```

instantiation.[7] The author of a model should provide a configuration unit that specifies how to assemble the model in question. The configuration unit becomes an integral part of the source of the model.

For example, as shown in Sample 7-39, there should be a configuration for building the behavioral model of a design, as well as a configuration for building the RTL model of a design, as shown in Sample 7-40. There should also be a configuration for building a model of a board.

Sample 7-39.
Configuration
for a behav-
ioral model

```
configuration beh_design of design is
for beh -- architecture of "design"
    ...
end for;
end beh_design;
```

Sample 7-40.
Configuration
for an RTL
model

```
configuration rtl_design of design is
for rtl   -- architecture of "design"
    ...
end for;
end rtl_design;
```

7. For more information on VHDL configuration, see page 273 of *VHDL Coding Styles and Methodologies*, 2nd edition, by Ben Cohen (Kluwer Academic Publisher)

Use configuration units for each testcase and test harness.

There should also be a configuration unit to specify the configuration of the test harness. An example is shown in Sample 7-41. Each testcase would also be configured using a configuration unit. The configuration of the testcase should include the configuration of test harness. Sample 7-42 shows an example of a testcase configuration unit.

Sample 7-41.
Configuration for a test harness

```
configuration main of harness is
for main   -- architecture of "harness"
    for cpu:cpu_server use ...
    end for;
end for;   -- architecture "main"
end main;
```

Sample 7-42.
Configuration for a testcase

```
library harness;
configuration testcase of bench is
for testcase  -- architecture of "bench"
    for th:harness use configuration harness.main;
    end for;    -- configuration "main"
    ...
end for;        -- architecture "testcase"
end testcase;
```

Generate the final configuration to select the design.

If you examine the configuration for the test harness and the testcase, you will notice it does not include a configuration specification for the design under verification. Default configuration cannot be used because you want the ability to change which model you want to simulate using a specific testcase. Once a model is compiled, the default configuration is already selected.

You could write a different configuration unit for each possible combination of testcase and design under verification. This would duplicate a lot of information and would be cumbersome to maintain should the structure of the testcase or harness be modified. It is easier to generate the final testcase configuration to include the configuration specification to the desired design.

Provide a configuration unit template.

The simulation script could parse the VHDL source files to automatically generate a configuration unit. However, it is best to provide a testcase configuration template to the script, with a clearly identifiable placeholder for the configuration specification for the design under verification. Sample 7-43 shows the testcase configuration unit modified to include a placeholder to be expanded by the simulation script. Using the template, the simulation script only needs to replace the placeholder with a reference to the desired design configuration, as shown in Sample 7-44. Notice how the final configuration unit is able to configure a component instantiation in an architecture that was previously configured in its own configuration unit. The generated configuration unit is compiled every time before each simulation.

Sample 7-43.
Configuration template for a test harness

```
library harness;
configuration testcase of bench is
for testcase  -- architecture of "bench"
    for th:harness use configuration harness.main;
    for main   -- architecture inside cfg "main"
        for duv:design
            use <design>;
        end for;
    end for;    -- architecture "main"
    end for;    -- configuration "main"
    . . .
end for;        -- architecture "testcase"
end testcase;
```

Sample 7-44.
Expanded configuration template for a test harness

```
library beh_lib;
library harness;
configuration testcase of bench is
for testcase  -- architecture of "bench"
    for th:harness use configuration harness.main;
    for main   -- architecture inside cfg "main"
        for duv:design
            use configuration beh_lib.beh_design;
        end for;
    end for;    -- architecture "main"
    end for;    -- configuration "main"
    . . .
end for;        -- architecture "testcase"
end testcase;
```

SDF Back-Annotation

SDF file are used to model accurate delays.

In a gate-level model, each gate is modeled using delays estimated from average output load conditions. However, in a real gate-level netlist, each gate is subject to different output loads: they drive different numbers of inputs, and the length of the wires connecting the output of the gate to the driven inputs are different. Each contributes to the load on the output of the gate, producing different loads for different instances of the same gate.

To be more accurate, gate-level simulations are back-annotated with delay values calculated from the physical netlist or the layout geometries. These more-accurate delay values are stored in a *Standard Delay File*. The *SDF* file is read by the simulator and each delay value replaces the average delay estimate for each instance. Thus, each gate instance can have a different delay value. The delay between an output pin and each of its driven input pins can also be different.

SDF annotation can take a long time.

Gate-level netlists can contain a few million gates and several million pin-to-pin nets or connections. Each must be annotated with a new delay value. This process can be very time-consuming and should be minimized whenever possible. If you have to recompile your model for each testcase, you have to perform the back-annotation each time as well.

Use compiled back-annotation whenever possible.

Compiled simulators usually offer compile-time back-annotation of a gate-level model. In that mode, the back-annotation of the delay values is performed once at compile time. Different testcases can be configured to run on the design in separate simulations without requiring that the back-annotation process be repeated.

Concatenate testcases to minimize back-annotation.

If you must use simulation-time back-annotation, you should minimize the time spent back-annotating the gate-level model of the design under verification. This can only be accomplished by minimizing the number of times the simulation is invoked. To invoke the simulation only once for multiple testcases, you need to concatenate each testcase into a single module or process that executes each in sequence.

Use a single procedure or task to control each testcase.

You still want each testcase to be written separately, and the ability to simulate them independently during development or when SDF back-annotation is not required. It is simple to concatenate each testcase into a sequence, if each testcase is encapsulated in a single *task* or *procedure*. To simulate a particular testcase, you simply have to call the *task* or *procedure* that encapsulates it. The testcases are concatenated by creating a sequence of *task* or *procedure* calls in a sequencing process.

In Verilog, add an *initial* block to execute the testcase stand-alone.

Sample 7-45 shows the Verilog testcase, originally shown in Sample 6-8 on page 234, encapsulated in a *task*. To execute this testcase in stand-alone mode, an *initial* block calls the task, if the simulation is invoked with the +*standalone* user-defined simulation-time option. To execute this testcase in stand-alone mode, presumably to debug a problem, you would compile only this testcase with the harness and the design under verification, as shown in Sample 7-46.

Sample 7-45.
Verilog
testcase encapsulated in a
task

```
module testcase;

task testcase;
    reg [  15:0] val;
    reg [64*8:1] msg;
begin
    harness.reset;
    harness.cpu.read(24'h00_FFFF, val);
    val[0] = 1'b1;
    harness.cpu.write(24'h00_FFFF, val);
    ...
    packet.send(msg);
end
endtask

initial
begin
    if ($test$plusargs("standalone")) begin
        testcase;
        syslog.terminate;
    end
end
endmodule
```

Sample 7-46.
Command-line
for standa-alone simula-tion

```
% mverilog -F tests/testcase.mft \
           -F beh/design.mft
           +standalone
```

Control the sequence of testcases using user-defined options.

The sequence of testcases is created in a separate module. It contains an *initial* block that can invoke all known testcases. To control which testcases are run and which ones are not, each *task* call is embedded in an *if* statement which tests for a user-defined command-line option. That way, you might run only a subset of testcases instead of all of them. Sample 7-47 shows the structure of the testcase sequencer module. To run a set of testcases, you simply have to specify the name of the desired testcases as a user-defined command-line option. To run all testcases, simply use the *+all_testcases* user-defined option. Sample 7-48 shows an example of each usage with a *VCS* compiled simulation. Notice how it was not necessary to recompile the model to execute different testcases.

Sample 7-47.
Sequencer
module for
simulating
multiple
testcases

```
module sequencer;
initial
begin: sequence
   reg all;

   all = $test$plusargs("all_testcases");

   if (all || $test$plusargs("testcase1") begin
       testcase1.testcase1;
   end
   if (all || $test$plusargs("testcase2") begin
       testcase2.testcase2;
   end
   ...
   syslog.terminate;
end
endmodule
```

Sample 7-48.
Sequencing
different
testcases

```
% vcs -F tests/all.mft -F gate/design.mft \
      -F phy/sdf.mft
% ./simv +testcase3 +testcase7
% ./simv +all_testcases
```

In VHDL, pass all client/server control signals to the encapsulating procedure.

Encapsulating the testcase in a VHDL procedure requires that all client/server control signals be passed through the procedure interface as *signal*-class arguments. This could be avoided if the procedure was declared in the same process where it will be used. This would require that all testcases, now encapsulated in procedures, be written in a single process declarative region. However, if you want to write and simulate each testcase independently, they will have to be located in packages, thus requiring the *signal*-class arguments. Sample 7-49 shows the testcase originally shown in Sample 6-19 on page 245 encapsulated in a procedure in a package.

Control the sequence of testcases using top-level generics.

The sequence of testcases is implemented in a process in the architecture instantiating the test harness. The process can invoke all known testcases. To control which testcases are run and which ones are not, each procedure call is embedded in an *if* statement checking if a top-level generic has been defined. That way, you might run only a subset of testcases instead of all of them.

Sample 7-49.
Encapsulated
client testcase

```
use work.i386sx;
package body testcase is

procedure do_testcase(
    signal to_srv : out i386sx.to_srv_typ;
    signal frm_srv: in  i386sx.frm_srv_typ)
is
    variable data: data_typ;
begin
    ...
    -- Perform a read
    i386sx.read(..., data, to_srv, frm_srv);
    ...
    -- Perform a write
    i386sx.write(..., ..., to_srv, frm_srv);
    ...
end do_testcase;
end testcase;
```

Sample 7-50 shows the structure of the testcase sequencer architecture. To run a set of testcases, you have to override the default value of the top-level generic corresponding to the desired testcase. The type of the top-level generics is *integer* because some VHDL simulators may not support user-defined or enumerated types when setting a generic from the command line. To run all testcases, simply override the value of the *all_testcases* top-level generic. Sample 7-51 shows how to select different testcase sequences by setting the top-level generics on the simulation command line under *Model-Sim*.

Output File Management

Simulations produce output files.

A simulation usually creates at least one output file. For example, *Verilog-XL* simulations generate a copy of the output messages in a file named "*verilog.log*" by default. Another frequently produced output file is the file containing the signal trace information for a waveform viewer. These output files are valuable. They are used to determine if the simulation was successful. They should be saved after each simulation run and parsed or post processed to determine success or failure.

Multiple simulations can clobber each other's files.

When you run only one simulation at a time, you can save them by renaming them after the completion of the simulation. That way, you can keep a history of testcases that were run on the design under verification. However, if you run multiple simulations in par-

Sample 7-50.
Sequencer
architecture
for simulating
multiple
testcases

```
entity sequencer;
    generic (all_testcases: natural := 0;
               testcase1   : natural := 0;
               testcase2   : natural := 0; ...);
end sequencer;

use work.testcase1;
use work.testcase2;
...
use work.i386sx;
...

architecture test of sequencer is
begin

    duv: design;

    run:process
    begin
        if all_testcases + testcase1 > 0 then
            testcase1.do_testcase1(i386sx.to_srv,
                                    i386sx.frm_srv);
        end if;
        if all_testcases + testcase2 > 0 then
            testcase2.do_testcase2(i386sx.to_srv,
                                    i386sx.frm_srv);
        end if;
        ...
        assert false
            report "Simulation completed"
            severity failure;
    end process run;
end test;
```

Sample 7-51.
Sequencing
different
testcases

```
% vsim -Gtestcase3=1 -Gtestcase7=1 sequencer
% vsim -Gall_testcases=1 sequencer
```

allel, usually on different machines, the output files from one simulation can clobber those of another. If you rely on default or hardcoded filenames, you will not be able to run simulations in parallel. You must be able to name files differently for different testcases.

Specify output filenames on the command line in your simulation run script.

A few default output filenames can be changed from the command line. For example, the *-l* option in Verilog can be used to change the name of the output *log* file. In "Configuration Management" on page 294, I recommended that you use a script to help manage the configuration of a simulation. That same script can also manage the naming of the output files according to the name of the testcase. Sample 7-52 shows how a PERL script can use the name of the testcase specified on the command line to rename the output log file in a Verilog simulation. Notice how it uses the manifest file parse script called *mverilog* to invoke the simulation with the proper set of files.

Sample 7-52.
Simulation run
script

```
require "getopts.pl";
&usage if &getopts("hr") || $opt_h || !@ARGV;

sub usage {
    print <<USAGE;
Usage: $0 [-r] {testcase}
  -r   Use the RTL model instead of behavioral
USAGE
    exit (1);
}

$design = ($opt_r) ? "rtl" : "beh";
$prefix  = "mverilog -F $design/design.mft ";

foreach $test (@ARGV) {
    $command = "$prefix -F tests/$test.mft";
    $command .= " -l logs/$test.log";
    system($command);
}
```

In VHDL, use a *string* type constant and generic.

Not all filenames can be renamed from the command line. In VHDL, all file objects are named from within the VHDL model. To make the filenames unique for each testcase, provide the name of the current testcase to the test harness to create unique filenames. In the testcase itself, you must be careful to use the testcase name to generate filenames as well. The simplest way is to have each testcase contain a constant defined as the (unique) testcase name. The value of that constant would be used to generate filenames and passed to the test harness via a generic. Sample 7-53 shows an example. Notice how the string concatenation operator is used to generate a unique filename in the file declarations.

Sample 7-53.
Generating
unique file
names in
VHDL

```
architecture test of bench is
    constant name: string := "testcase";
    component harness
        generic (name: string);
    end component;
begin

    th: harness generic map (name);

    process
        file results: text is out name & ".out";
    begin
        . . .
    end process;
end test;
```

In Verilog, use a
parameter.

A similar technique can be used in Verilog. You simply use a
parameter instead of the constant and pass the testcase name to the
test harness as a parameter association. Strings in Verilog are simply bit vectors with eight bits per character. They can be concatenated using the usual concatenation operator to create unique
filenames. Sample 7-54 shows an example.

Sample 7-54.
Generating
unique file
names in Ver-
ilog

```
module testbench is
parameter name = "testcase";

harness #(name) th ();

initial
begin
    integer results;
    results = $fopen({name, ".out"});
    . . .
end
endmodule
```

REGRESSION

A regression
ensures backward
compatibility.

A regression suite ensures that modifications to the design remain
backward compatible with previously verified functionality. Many
times, a change in the design made to fix a problem detected by a
testcase, will break functionality that was previously verified. Once
a testbench is written and debugged to simulate successfully, you
must make sure that it continues to be succesful throughout the
duration of the design project.

Running Regressions

Regressions are run at regular intervals.

As individual self-checking testbenches are completed, they are added to a master list of testcases included in the regression simulation. This regression simulation is run at regular intervals, usually nightly. As the number of testcases in the regression suite grows, it may not be possible to complete a full regression simulation overnight. Testcases can then be classified into two groups: one group is run every night, while the second group is only included in regressions run over a week end.

Testbenches may have a *fast* mode to speed-up regressions.

Another approach is to provide a *fast mode* to testcases where only a subset of the functionality is verified during overnight regression simulations. The full functionality would be verified only during invidual simulations or regression simulations over a week end. In Verilog, the fast mode could be turned on using a user-defined command-line option, as shown in Sample 7-55. In VHDL, it could be turned on using a top-level generic, as shown in Sample 7-56.

Sample 7-55.
Implementing a fast mode in a Verilog testbench

```
% verilog ... +fastmode

module testcase;
...
initial
begin
    // Repeat only 4 times in fast mode
    repeat (($test$plusarg("+fastmode"))?4:256)
        begin
            ...
        end
    syslog.terminate;
end
endmodule
```

Use a script to run regressions.

A regression script could invoke each testcase in the regression test suite using the simulation configuration script used to invoke individual simulations, as discussed in "Configuration Management" on page 294. If the number and duration of testcases in the regression suite make it impossible to run a regression simulation in the allotted time, you will want to consider parallel simulations. If you do, it is necessary that testbenches be designed to produce results independently from each other, as discussed in "Output File Management" on page 309. Parallel simulations can be managed using readily available utilities, such as *pmake*, *Load Balancer*, or *LSF*.

Sample 7-56.
Implementing
a fast mode in
a VHDL test-
bench

```
entity bench is
    generic (fast_mode: integer := 0);
end bench;

architecture test of bench is
begin
    process
        variable repeat: integer := 256;
    begin
        -- Repeat only 4 times in fast mode
        if fast_mode /= 0 then
            repeat := 4;
        end if;
        for I in 1 to repeat loop
            ...
        end loop;
        assert false
            report "Simulation completed"
            severity failure;
    end process;
end test;
```

Regression Management

Check out a fresh
view with local
copies.

Not all source files are suitable for regression runs. If you are using your revision control system properly, you should be checking in files at times convenient for you, not convenient for the regression run. The latest version of a file might contain code that was not tested at all or that might even have syntax errors. You do not want to waste a regression simulation on files that were not properly debugged. Before running a regression, you should check-out a complete view of the source control database, populated with local copies whose revisions are tagged as being suitable for regression testing. This tag is applied by verification and design engineers once they have confidence in the basic functionality of the code and are ready to submit that particular revision of the testcase or the design to regression. Sample 7-57 shows how to tag a particular revision of a file, then check-out the particular revision of a file associated with a tag in RCS. More advanced revision control systems allow tagging and checking-out entire file systems.

Put a timebomb in
all simulations.

One of the greatest killers of regression simulations, second only to the infinite loop, is the simulation that never terminates. A simulation will run forever if a condition you are waiting for never occurs. The clock generator keeps the simulation alive by continuously

Sample 7-57.
Tagging and
retrieving a
particular revi-
sion of a file

```
% rcs -r1.6 -nregress: design.vhd
...
% co -rregress *.vhd
```

generating events. Time advances until the maximum value is reached, which, in modern simulators using 64-bit time values, will take a long time! To prevent a testcase from hanging a regression simulation, include a timebomb in all simulations. This timebomb should go off after a delay long enough to allow the normal operations of the testcase to complete without interruption. Sample 7-58 shows a timebomb, used with a concurrent procedure call in Sample 7-59. The procedure could be modified to include a signal argument that, when triggered with an event, would reset the fuse.

Sample 7-58.
Timebomb
procedure

```
package body bomb_pkg is

procedure timebomb(constant fuse: in time) is
begin
    wait for fuse;
    assert false
        report "Boom!"
        severity failure;
end timebomb;

end bomb_pkg;
```

Sample 7-59.
Using the
timebomb pro-
cedure

```
use work.bomb_pkg.all;
architecture test of bench is
begin
    bomb: timebomb(fuse => 12 ms);

    test_procedure: process
    begin
        ...
    end process test_procedure;
end test;
```

Do not rely on a
timebomb for nor-
mal termination.

The timebomb should only be used to prevent run-away simulations from running forever. It should not be used to terminate a testcase under normal conditions. It would be impossible to distinguish between a successful completion of the testcase and a dead-lock condition. Furthermore, the timebomb would require fine tuning everytime the testbench or design is modified to avoid the testcase from being prematurely interrupted or wasting simulation cycles by running for too long.

Automatically generate a report after each regression run.

Once the regression simulation is completed, the success or failure of each testcase in the regression suite should be checked using the output log scan script (see "Pass or Fail?" on page 289.) The results are then summarized into a single regression report outlining which particular testcase was successful or failed. It is a good idea to have the regression script mail the report to all the engineers in the design team to ensure that the design remains backward compatible at all times. This report should also be the first item on the agenda in any design team meeting.

SUMMARY

This chapter described how behavioral models can be used to accelerate the verification of a design project by improving simulation performance and parallelizing the verification effort. Behavioral models let system-level verification start sooner and be performed using regular tools and hardware platforms. To obtain all of these benefits from behavioral models, they must be written using skills and approaches different from RTL models.

This chapter showed how to manage simulations: from determining success or failure to configuration management to regression simulation. It is recommended that the output of simulation runs be parsed to determine if the response was as expected. The use of a configuration script is also recommended to remove the verification engineers from the details of a model structure and composition.

APPENDIX A   CODING GUIDELINES

There have been many sets of coding guidelines published for hardware description languages, but they have historically focused on the synthesizable subset and the target hardware structures. Writing testbenches involves writing a lot of code and also requires coding guidelines. These guidelines are designed to enhance code maintainability and readability, as well as to prevent common or obscure mistakes.

Guidelines are
structured from
the generic to
the specific.

The guidelines presented here are reproduced with permission from the *Reuse Methodology Field Guide* from Qualis Design Corporation (http://www.qualis.com). They are organized from the general to the specific. They start with general coding guidelines that should be used in any language. They are followed by guidelines specific to hardware description languages. Verilog and VHDL-specific guidelines follow after that. Note: a guideline applicable to a more specific context can contradict and supersede a more general guideline.

Define guide-
lines as a group,
then follow
them.

Coding guidelines have no functional benefits. Their primary contribution is toward creating a readable and maintainable design. Having common design guidelines makes code familiar to anyone familiar with the implied style, regardless of who wrote it. The primary obstacle to coding guidelines are personal preferences. It is important that the obstacle be recognized for what it is: personal taste. There is no intrinsic value to a particular set of guidelines. The value is in the fact that these guidelines are shared by the entire group. If even one individual does not follow them, the entire group is diminished.

DIRECTORY STRUCTURE

Use an identical directory structure for every project.

Using a common directory structure makes it easier to locate design components and to write scripts that are portable from one engineer's environment to another. Reusable components that were designed using a similar structure will be more easily inserted into a new project.

Example project-level structure:

```
.../bin/        Project-wide scripts/commands
   doc/         System-level specification documents
   SoCs/        Data for SoCs/ASICs/FPGA designs
   boards/      Data for board designs
   systems/     Data for system designs
   mech/        Data for mechanical designs
   shared/      Data for shared components
```

At the project level, there are directories that contain data for all design components for the project. Components shared, unmodified, among SoC/ASIC/FPGA, board, and system designs are located in a separate directory to indicate that they impact more than a single design. At the project level, shared components are usually verification and interface models.

Some "system" designs may not have a physical correspondence and may be a collection of other designs (SoCs, ASICs, FPGAs, and boards) artifically connected together to verify a subset of the system-level functionality.

Each design in the project has a similar structure. Example of a design structure for an SoC:

```
SoCs/name/       Data for ASIC named "name"
      doc/       Specification documents
      bin/       Scripts specific to this design
      beh/       Behavioral model
```

rtl/	Synthesizable model
syn/	Synthesis scripts & logs
phy/	Physical model and SDF data
verif/	Verif suite and simulation logs
SoCs/shared/	Data for shared ASIC components

Components shared, unmodified, between SoC designs are located in a separate directory to indicate that they impact more than a single design. At the SoC level, shared components include processor cores, soft and hard IP, and internally reused blocks.

Use relative pathnames.

Using absolute pathnames requires that future use of a component or a design be installed at the same location. Absolute pathnames also require that all workstations involved in the design have the design structure mounted at the same point in their file systems. The name used may no longer be meaningful and the proper mount point may not be available.

If full pathnames are required, use preprocessing or environment variables.

Put a Makefile with a default 'all' target in every source directory.

Makefiles facilitate the compilation, maintenance, and installation of a design or model. With a Makefile the user need not know how to build or compile a design to invoke "make all". Top-level makefiles should invoke *make* on lower level directories.

Example "all" makefile rule:

```
all: subdirs ...

SUBDIRS = ...
subdirs:
      for subdir in $(SUBDIRS); do \
         (cd $$subdir; make); \
      done
```

VHDL Specific

Locate the directories containing the libraries as subdirectories of the source directories.

It is a good idea to name a library according to the VHDL toolset it corresponds to. This naming convention makes it possible to use more than one VHDL toolset on the same source installation.

Example:

```
.../SoC/name/beh/
                work.nc/     NC library
                work.msim/   ModelSim library
                work.vss/    VSS library
            rtl/
                work.nc/     NC library
                work.msim/   ModelSim library
                work.vss/    VSS library
```

Create a file that lists all required libraries (other than WORK) and lists the full relative path name to the directory containing the source files for that library.

Having this file makes it easier to locate all source files required by a design or a portion of a design. This file can also be processed by a script that automatically generates the VHDL toolset library map file, which associates logical library names and container directories.

Verilog Specific

Create a file that lists all required source files and command-line options for simulating the design in every directory that contains a Verilog description of a design (or sub-design).

The file, called a manifest file, is used with the -f option when invoking the Verilog simulator. Using a front-end script, such the one described in "Verilog Configuration Management" on page 295, lets relative pathnames work reguardless of current working directory where the simulation is invoked. Manifest files with front-end scripts make configuration management more portable. A front-end script can also handles the case where a file is included more than once in a simulation through separate references in different manifest files.

To include a sub-design in a higher-level design, include the manifest file for the sub-design using the -f option in the higher-level manifest file.

This structure effectively creates hierarchical manifest files.

Specify files loaded using the `include directive using a complete relative pathname.

Requiring the use of a +*incdir* option on the Verilog command line makes it impossible to determine, from the source code only, which files are required to completely describe a model. The exact command line used is also required to reproduce any problems. If a complete relative pathname is specified, it becomes easy to locate all files required to make up a complete model.

GENERAL CODING GUIDELINES

These guidelines are intended to be used for any programming or scripting language. Additional guidelines for HDL and language-specific descriptions can be found in the section titled "HDL Coding Guidelines" on page 336.

Comments

Put the following information into a header comment for each source file: copyright notice, brief description, revision number, original author name and contact data, and current author name and contact data.

Example (PERL script under RCS):

```
#! /usr/local/bin/perl
#
# (c) Copyright 1999, Qualis Design Corporation
# All rights reserved.
#
# This file contains proprietary and confidential
# information. The content or any derivative work
# can only be used by or distributed to licensed
# users or owners.
#
# Description:
```

```
#      This script parses the output of a set of
#      simulation log files and produces a
#      regression report.
#
# Original author: John Q. Doe <jdoe@qualis.com>
# Current author : Jane D. Hall <jhall@qualis.com>
# Revision        : $Revision$
#
```

Use a trailer comment describing revision history for each source file.

The revision history should be maintained automatically by the source management software. Because these can become quite lengthy, put revision history at the bottom of the file. This location eliminates the need to wade through dozens of lines before seeing the actual code.

Example (shell script under RCS):

```
#
# History:
#
# $Log$
#
```

Use comments to describe the intent or functionality of the code, not its behavior.

Comments should be written with the target audience in mind: a junior engineer who knows the language, but is not familiar with the design, and must maintain and modify the design without the benefit of the original designer's help.

Example of a useless comment (C):

```
/* Increment i */
i++;
```

Example of a useful comment (C):

```
/
* Move the read pointer to the next input element
*/
i++;
```

Preface each major section with a comment describing what it does, why it exists, how it works, and assumptions on the environment.

> A major section could be a *process* in VHDL, an *always* block in Verilog, a *function* in C or a long sequence of code in any language.
>
> It should be possible to understand how a piece of code works by looking at the comments only and by stripping out the source code itself. Ideally, the source stripped of comments should be just as understandable.

Describe each input and output of subprograms in individual comments.

> Describe the purpose, expected valid range, and effects of each input and output of all subprograms or other coding unit supported by the language. Whenever possible, show a typical usage.
>
> Example (PERL):

```
#
# Subroutine to locate all files matching a given
# pattern under a given directory path
#
sub scandir {        # Returns array of filenames
    local($dir,      # Dir to recursively scan (str)
          $pattern)  # Filename pattern (regexp str)
          = @_;

   . . .
}
```

Delete bad code; do not comment-out bad code.

> Commented-out code begs to be reintroduced. Use a proper revision control system to maintain a track record of changes.

Layout

Use a minimum of three spaces for each identation level.

> An indentation that is too narrow (such as 2), does not allow for easily identifying nested scopes. An identation level that is too wide (such as 8), quickly causes the source code to reach the right margin.

Write only one statement per line.

The human eye is trained to look for sequences in a top-down fashion, not down-and-sideways. This layout also gives a better opportunity for comments.

Example of poor code layout (PERL):

```
$| = 1; print "Continue (y/n) ? [y] ";
$ans = <STDIN>; last if $ans =~ m/^\s*[nN]/;|
```

Example of good code layout (PERL):

```
# Prompt the user...
$| = 1;
print "Continue (y/n) ? [y] ";

# Read the answer
$ans = <STDIN>;

# Get out if answer started with a "n" or "N"
last if $ans =~ m/^\s*[nN]/;
```

Limit line length to 72 characters. If you must break a line, break it at a convenient location with the continuation statement and align the line properly within the context of the first token.

Printing devices are still limited to 80 characters in width. If a fixed-width font is used, most text windows are configured to display up to 80 characters on each line. Relying on the automatic line wrapping of the display device may yield unpredictable results and unreadable code.

Example of poor code layout (Verilog):

```
#10
expect = $realtobits((coefficient * datum)
        + 0.5);
```

Example of good code layout (Verilog):

```
#10 expect = $realtobits((coefficient * datum)
                          + 0.5);
```

Use a tabular layout for lexical elements in consecutive declarations, with a single declaration per line.

A tabular layout makes it easier to quickly scan the list of declarations, identifying their types, classes, initial values, etc... If you use a single declaration per line, it is easier to locate a particular declaration. A tabular layout also facilitates adding and removing a declaration.

Example of poor declaration layout (VHDL):

```
signal counta, countb: integer;
signal c: real := 0.0;
signal sum: signed(0 to 31);
signal z: unsigned(6 downto 0);
```

Example of good declaration layout (VHDL):

```
signal counta: integer;
signal countb: integer;
signal c     : real                    := 0.0;
signal sum   : signed   (0 to    31);
signal z     : unsigned (6 downto 0 );
```

If supported by the language, use named associations when calling subprograms or instantiating subunits. Use a tabular layout for lexical elements in consecutive associations, with a single association per line.

Using named associations is more robust than using port order. Named associations do not require any change when the argument list of a subprogram or subunit is modified or reordered. Furthermore, named associations provide for self-documenting code as it is unnecessary to refer to another section of the program to identify what value is being passed to which argument. A tabular layout makes it easier to quickly scan the list of arguments being passed to a subprogram. If you use one named association per line, it is easier to locate a particular association. A tabular layout also facilitates adding and removing arguments.

Example of poor association layout (Verilog):

```
fifo in_buffer(voice_sample_retimed,
               valid_voice_sample, overflow, ,
               voice_sample, 1'b1, clk_8kHz,
               clk_20MHz);
```

Example of good association layout (Verilog):

```
fifo in_buffer(.data_in   (voice_sample),
               .valid_in  (1'b1),
               .clk_in    (clk_8kHz),
               .data_out  (voice_sample_retimed),
               .valid_out (valid_voice_sample),
               .clk_out   (clk_20MHz),
               .full      (overflow),
               .empty     ());
```

Syntax

Do not use abbreviations.

Some languages, particularly scripting languages, allow using an abbreviated syntax, usually as long as the identifiers are unique prefixes. Long form and command names are self-documenting and provide a more consistent syntax than various abbreviations. If additional commands are later added to the language, abbreviations that used to be unique may now conflict with the new commands and require modification to remain compatible with the newer versions.

Example of poor command syntax (DC-shell):

```
re -f verilog design.v
```

Example of good command syntax (DC-shell):

```
read -format verilog design.v
```

Encapsulate repeatedly used operations or statements in subprograms.

By using subprograms, maintainance is reduced significantly. Code only needs to be commented once and bugs only need to be fixed once. It also reduces code volume.

Example of poor expression usage (Verilog):

```
// sign-extend both operands from 8 to 16 bits
operand1 = {{8 {ls_byte[7]}}, ls_byte};
operand2 = {{8 {ms_byte[7]}}, ms_byte};
```

Example of proper use of subprogram (Verilog):

```
// sign-extend an 8-bit value to a 16-bit value
function [15:0] sign_extend;
   input [7:0] value;
   sign_extend = {{8 {value[7]}}, value};
endfunction

//= sign-extend both operands from 8 to 16 bits
operand1 = sign_extend(ls_byte);
operand2 = sign_extend(ms_byte);
```

Use a maximum of 50 consecutive sequential statements in any statement block.

Too many statements in a block create many different possible paths. This makes it difficult to grasp all of the possible implications. It may be difficult to use a code coverage tool with a large statement block. A large block may be broken down using subprograms.

Use no more than three nesting levels of control-flow statements.

Understanding the possible paths through several levels of control flow becomes exponentially difficult. Too many levels of decision-making may be an indication of a poor choice in processing sequence or algorithm. Break up complex decision structures into separate subprograms.

Example of poor control-flow structure (C):

```
if (a == 1 && b == 0) {
    switch (val) {
```

```
   4:
   5: while (!done) {
          if (val % 2) {
              odd = 1;
              if (choice == val) {
                  for (j = 0; j < val; j++) {
                      select[j] = 1;
                  }
                  done = 1;
              }
          } else {
              odd = 0;
          }
      }
      break;
   0: for (i = 0; i < 7; i++) {
          select[j] = 0;
      }
      break;
   default:
        z = 0;
   }
}
```

Example of good control-flow structure (C):

```
void
process_selection(int val)
{
   odd = 0;
   while (!done) {
      if (val % 2) {
         odd = 1;
      }
      if (odd && choice == val) {
         for (j = 0; j < val; j++) {
            select[j] = 1;
         }
         done = 1;
      }
   }
}

if (a == 1 && b == 0) {
```

```
switch (val) {
0: for (i = 0; i < 7; i++) {
      select[j] = 0;
   }
   break;
4:
5: process_selection(val);
   break;
default:
   z = 0;
}
}
```

Debugging

Include a mechanism to automatically exclude all debug statements.

Debug information should be excluded by default and should be enabled automatically via a control file or command line options. Do not comment out debug statements and then uncomment them when debugging. This requires significant editing. When available, use a preprocessor to achieve better runtime performance.

Example of poor debug statement exclusion (Verilog):

```
// $write("Address = %h, Data = %d\n",
//         address, data);
```

Example of proper debug statement exclusion (Verilog):

```
`ifdef DEBUG
   $write("Address = %h, Data = %d\n",
           address, data);
`endif
```

NAMING GUIDELINES

These guidelines suggest how to select names for various user-defined objects and declarations. Additional restrictions on naming can be introduced by more specific requirements.

Capitalization

Use lowercase letters for all user-defined identifiers.

> Using lowercase letters reduces fatigue and stress from constantly holding down the Shift key. Reserve uppercase letters for identifiers representing special functions.

Do not rely on case mix for uniqueness of user-defined identifiers.

> The source code may eventually be processed by a case-insensitive tool. The identifiers would then lose their uniqueness. Use naming to differentiate identifiers.

Example of bad choice for identifier (C):

```
typedef struct RGB {
    char red;
    char green;
    char blue;
} RGB;

main () {
    RGB rgb;
    . . .
}
```

Example of better choice for identifier (C):

```
typedef struct rgb_struct {
    char red;
    char green;
    char blue;
} rgb_typ;

main () {
    rgb_typ rgb;
    . . .
}
```

In a case-insensitive language, do not rely on case mix for adding semantic to identifiers.

Instead of using the case of identifiers to document variable types, use naming (prefix, suffix) to self-document identifiers. Consistent case usage for a given identifier cannot be enforced by the compiler and therefore may end up being used incorrectly.

Example of poor choice of identifier (VHDL):

```
package Pci is
    type command is (MEM, IO, CONFIG);
    procedure readCycle(ADDRESS: in ...;
                        data: out ...);
end Pci;
```

Example of proper choice of indentifier (VHDL):

```
package pci_pkg is
    type command_typ is (MEM, IO, CONFIG);
    proceedure read_cycle(address_in: in ...;
                          data_out: out ...);
end pci_pkg;
```

Use suffixes to semantically differentiate related identifiers.

The suffix could indicate object kind, such as: type, constant, signal, variable, flip-flop, etc... or the suffix could indicate pipeline processing stage or clock domains.

Use uppercase letters for constant identifiers (run-time or compile-time).

The case differentiates between a symbolic literal value and a variable.

Example (Verilog):

```
module block(...);
...
`define DEBUG
parameter WIDTH = 4;
...
endmodule
```

Separate words using an underscore; do not separate words by mixing upper-case and lowercase letters

It can be difficult to read variables that use case to separate word boundaries. Using spacing between words is more natural. In a case-insensitive language or if the code is processed through a case-insensitive tool, the case convention cannot be enforced by the compiler.

Example of poor word separation (C):

```
readIndexInTable = 0;
```

Example of proper word separation (C):

```
read_index_in_table = 0;
```

Identifiers

Do not use reserved words of popular languages or languages used in the design process as user-defined identifiers.

Not using reserved words as identifiers avoids having to rename an object to a synthetic, often meaningless, name when translating or generating a design into another language. Popular languages to consider are C, C++, Verilog, VHDL, PERL, VERA, and e.

Use meaningful names for user-defined identifiers and use a minimum of five characters.

Avoid acronyms or meaningless names. Using at least five characters increases the likelyhood of using full words.

Example of poor identifier naming (VHDL):

```
if e = '1' then
    c := c + 1;
end if;
```

Example of good identifier naming (VHDL):

```
if enable = '1' then
   count := count + 1;
end if;
```

Name objects according to function or purpose; avoid naming objects according to type or implementation.

This naming convention produces more meaningful names and automatically differentiates between similar objects with different purposes.

Example of poor identifier naming (Verilog):

```
count8 <= count8 + 8'h01;
```

Example of good identifier naming (Verilog):

```
addr_count <= addr_count + 8'h01;
```

Do not use identifiers that are overloaded or hidden by identical declarations in a different scope.

If the same identifier is reused in different scopes, it may become difficult to understand which object is being referred to.

Example of identifier overloading (Verilog):

```
reg [7:0] address;

begin: decode
   integer address;

   address = 0;
   ...
end
```

Example of good identifier naming (Verilog):

```
reg [7:0] address;

begin: decode
   integer decoded_address;

   decoded_address = 0;
   ...
end
```

Constants

Use symbolic constants instead of "magic" hard-coded numeric values.

Numeric values have no meaning in and of themselves. Symbolic constants add meaning and are easier to change globally. This is especially true if several constants have an identical value but a different meaning. If the language does not support symbolic constants, use a preprocessor or a variable appropriately named.

Example of poor constant usage (C):

```
int table[256];

for (i = 0; i <= 255; i++) ...
```

Example of good constant usage (C):

```
#define TABLE_LENGTH 256

int table[TABLE_LENGTH];

for (i = 0; i < TABLE_LENGTH; i++) ...
```

HDL Specific

Number multi-bit objects using the range N:0.

Using this numbering range avoids accidental truncation of the top bits when assigned to a smaller object (Verilog). This convention also provides for a consistent way of accessing bits from a given direction. If the object carries an integer value, the bit number represents the power-of-2 for which this bit corresponds.

Example (Verilog):

```
parameter width = 16;

reg [     7:0] byte;
reg [    31:0] dword;
reg [width-1:0] data;
```

Example (VHDL):

```
generic(width: integer := 16);
...
variable byte : unsigned (      7 downto 0);
variable dword: unsigned (     31 downto 0);
variable data : unsigned (width-1 downto 0);
```

Do not specify a bit range when referring to a complete vector.

If the range of a vector is modified, all references would need to be changed to reflect the new size of the vector. Using bit ranges implicitly means that you are referring to a subset of a vector. If you want to refer to the entire vector, do not specify a bit range.

Example of poor vector reference (VHDL):

```
signal count: unsigned(15 downto 0);
...
count(15 downto 0) <= count(15 downto 0) + 1;
carry <= count(15);
```

Example of proper vector reference (VHDL):

```
signal count: unsigned(15 downto 0);
...
count <= count + 1;
carry <= count(count'left);
```

Preserve names across hierarchical boundaries.

Preserving names across hierarchical boundaries facilitates tracing signals up and down a complex design hierarchy. This naming convention is not applicable when a unit is instantiated more than once, or when the unit was not originally designed within the context of the current design.

Filenames

Use filename extensions that indicate the content of the file.

Tools often switch to the proper language-sensitive function based on the filename extension. Use a postfix on the filename itself if different (but related) contents in the same language are provided. Using postfixes with identical root names causes all related files to show up next to each other when looking up the content of a directory.

Example of poor file naming (Verilog):

```
design.vt          Testbench
design.vb          Behavioral model
design.vr          RTL model
design.vg          Gate-level model
```

Example of good file naming (Verilog):

```
design_tb.v        Testbench
design_beh.v       Behavioral model
design_rtl.v       RTL model
design_gate.v      Gate-level model
```

HDL CODING GUIDELINES

The following guidelines are specific to HDL descriptions. These guidelines are presented in addition to the guidelines outlined for general coding and naming. Additional guidelines will be needed when describing a design to be synthesized.

Structure

Use a single compilation unit in a file.

> A file should contain a single module (Verilog), or a single entity, architecture, package, package body, or configuration (VHDL). This structure facilitates locating various model components. For VHDL, it further reduces the amount of recompilation that may be required.

Layout

Declare ports and arguments in logical order according to purpose or functionality; do not declare ports and arguments according to direction.

> Group port declarations that belong to the same interface. Grouping port declarations facilitates locating and changing them to a new interface. Do not order declarations output first, data input second, and control signals last because it scatters related declarations.

Lay out code for maximum readability and maintainability.

> Saving a few lines or characters does not save money. Saving 1 character on a $200 1G disk saves 0.00005 cents. However, saving 1 minute of an engineer's (or your own) time trying to understand your code saves between $.50 and $1.

VHDL Specific

Label all processes, concurrent statements and loops.

> Labeling makes referring to a particular construct much easier. If a particular construct is not named, some features in debugging tools may not be available. Labeling also provides for an additional opportunity to document the code.
>
> Example of a named loop:

```
scan_bits_lp: for i in data'range loop
   ...
   exit scan_bits_lp when data(i) = 'X';
end loop scan_bits_lp;
```

Example of a named process:

```
clock_generator: process
begin
    wait for 50 ns;
    CLK <= not CLK;
end process clock_generator;
```

Label closing "end" keywords.

The "begin" and "end" keywords may be separated by hundreds of lines. Labeling matching "end" keywords facilitates recognizing the end of a particular construct. If the VHDL-87 syntax does not support a labeled "end" keyword, add the label using a comment.

Example:

```
component FIFO
    generic (...);
    port (...);
end component; -- FIFO
```

Use inline range constraints instead of subtypes.

Because type and subtype names are not syntactically different, using too many subtypes makes it hard to remember what type remains compatible with what other type.

Example of subtype constraints:

```
subtype address_styp is
    std_logic_vector (15 downto 0);
subtype count_styp   is
    integer range 15 downto 0;

signal address: address_styp;
signal count   : count_styp;
```

Writing Testbenches: Functional Verification of HDL Models

Example of inline range constraints:

```
signal address: std_logic_vector (15 downto 0);
signal count   : integer        range 0 to 15;
```

Do not use ports of mode "*buffer*" and "*linkage*".

Buffer ports have special requirements when instantiated in a higher level unit. Use an "out" port instead. If internal feedback is required, use an internal feedback signal. I am still not sure what linkage ports were designed for. Coolant fluid?

Example using an internal feedback signal:

```
port(was_buffer_mode: out std_logic);
...
signal was_buffer_mode_int: std_logic;
...
was_buffer_mode <= was_buffer_mode_int;
```

Do not use blocks with ports and generics.

Ports and generics on blocks can be used to rename signals and constants already visible, thus creating a second name for an object. Using ports and generics on blocks reduces maintainability. Use blocks only when a local declarative region is required (e.g. to configure instantiations in a generate statement or declare an intermediate signal).

Do not use guarded expressions, signals and assignments, driver disconnect and signal kinds "*bus*" and "*register*".

These features are scheduled to be removed from the language. Furthermore, they are used so little that tools may be unreliable when using them.

Use the logical library name "WORK" when referring to units in the same library.

Using this logical name makes it possible for a design to be moved or copied into another library with a different name without requiring any modifications. It also eliminates the need for a particular library name to hold the design.

Using "*WORK*" is similar to using the relative directory name "." whereas using the actual library name is similar to using a full pathname. The former is portable to a different environment. The latter is not.

Verilog Specific

Start every module with a `resetall directive.

Compiler directives remain active across file boundaries. A module inherits the directives defined in earlier files. This may create compilation-order dependencies in your model. Using the `resetall directive ensures that your module is not affected by previously-defined compiler directives and will be properly self-contained.

Make sure your module name is unique.

Module names are global to the compilation and may interfere with other module names in the same simulation. Use a prefix that is unique to your verification environment or your design to make sure the module name you choose will be unique.

Example of poor module naming:

```
module cpuif(...);
...
endmodule
```

Example of proper module naming:

```
module xyz_cpuif(...);
...
endmodule
```

Avoid using `define symbols.

`define symbols are global to the compilation and may interfere with other symbols defined in another source file. For constant values, use parameters. If `define symbols must be used, undefine them by using `undef when they are no longer needed.

Example of poor style using `define symbols:

```
`define CYCLE 100
`define ns      * 1
always
begin
   #(`CYCLE/2 `ns);
   clk = ~clk;
end
```

Example of good style avoiding `define symbols:

```
parameter CYCLE = 100;
`define ns * 1
always
begin
   #(CYCLE/2 `ns);
   clk = ~clk;
end
`undef ns
```

Use a non-blocking assignment for registers used outside the *always* or *initial* block where the register was assigned.

Using non-blocking assignments prevent race conditions between blocks that read the current value of the reg and the block that updates the reg value. This assignment guarantees that simulation results will be the same across Verilog simulators or with different command-line options.

Example of coding creating race conditions:

```
always @ (s)
begin
   if (s) q = q + 1;
end

always @ (s)
begin
   $write("Q = %b\n", q);
end
```

Example of good portable code:

```
always @ (s)
begin
    if (s) q <= q + 1;
end

always @ (s)
begin
    $write("Q = %b\n", q);
end
```

Assign regs from a single *always* or *initial* block.

Assigning regs from a single block prevents race conditions between blocks that may be setting a reg to different values at the same time. This assignment convention guarantees that simulation results will be the same across Verilog simulators or with different command-line options.

Example of coding that creates race conditions:

```
always @ (s)
begin
    if (s) q <= 1;
end

always @ (r)
begin
    if (r) q <= 0;
end
```

Example of good portable code:

```
always @ (s or r)
begin
    if (s)      q <= 1;
    else if (r) q <= 0;
end
```

Do not disable tasks with output or inout arguments.

The return value of output or inout arguments of a task that is disabled is not specified in the Verilog standard. Disable the inner begin/end block instead of disabling tasks with ouput or inout arguments. This guarantees that simulation results will be the same across Verilog simulators or with different command-line options.

Example of coding with unspecified behavior:

```
task cpu_read;
   output [15:0] rdat;
begin
   . . .
   if (data_rdy) begin
      rdat = data;
      disable cpu_read;
   end
   . . .
end
endtask
```

Example of good portable code:

```
task cpu_read;
   output [15:0] rdat;
begin: do_read
   . . .
   if (data_rdy) begin
      rdat = data;
      disable do_read;
   end
   . . .
end
endtask
```

Do not disable blocks containing non-blocking assignments with delays.

What happens to pending non-blocking assignments performed in a disabled block is not specified in the Verilog standard. Not disabling this type of block guarantees that simulation results will be the same across Verilog simulators or with different command-line options.

Example of coding with unspecified behavior:

```
begin: drive
    addr <= #10 16'hZZZZ;
    ...
end

always @ (rst)
begin
    if (rst) disable drive;
end
```

Do not read a wire after updating a register in the right-hand side of a continuous assignment, after a delay equal to the delay of the continuous assignment.

The Verilog standard does not specify the order of execution when the right-hand side of a continuous assignment is updated. The continuous assignment may be evaluated at the same time as the assignment, or in the next delta cycle.

If you read the driven wire after a delay equal to the delay of the continuous assignment, a race condition will occur. The wire may or may not have been updated.

Example creating a race condition:

```
assign qb = ~q;
assign #5 qq = q;
initial
begin
    q = 1'b0;
    $write("Qb = %b\n", qb);
    #5;
    $write("QQ = %b\n", qq);
end
```

Do not use the bitwise operators in a boolean context.

Bitwise operators are for operating on bits. Boolean operators are for operating on boolean values. They are not always interchangeable and may yield different results. Use the bitwise operators to indicate that you are operating on bits, not for making a decision based on the value of particular bits.

Some code coverage tools cannot interpret a bitwise operator as a logical operator and will not provide coverage on the various components of the conditions that caused the execution to take a particular branch.

Example of poor use of bitwise operator:

```
reg [3:0] BYTE;
reg VALID
if (BYTE & VALID) begin
   ...
end
```

Example of good use of boolean operator:

```
reg [3:0] BYTE;
reg VALID
if (BYTE != 4'b0000 && VALID) begin
   ...
end
```

AFTERWORDS

This book should have given you the necessary skills to plan, implement and manage a best-in-class verification process. The methodologies and techniques will need to be tuned to your specific requirements. Think of this book as providing you with a set of *Lego* blocks. It is now up to you to put them together to build the infrastructure you envision.

Training classes are available.

If you would like to attend a formal training class covering the techniques presented in this book, I recommend the language and methodology classes[1] offered by *Qualis Design Corporation* (www.qualis.com). Just like this book, they focus on the methodology and how to implement it efficiently, not the tools. These classes are taught by professional engineering consultants (sometimes by myself) who spend most of their time applying these techniques on leading-edge designs. They draw on their extensive industry experience to answer any question you may have on verification, adapting the techniques to your circumstances, often going beyond the content of the class material.

Join the on-line *verification guild.*

Send me email at janick@bergeron.com and ask to be added to the *verification guild* mailing list. It is a moderated on-line forum to discuss verification-related issues. Verification languages, behavioral modeling, testbench structures, detailed syntax of a waveform

---

1. Of course I am going to recommend them: I wrote the bulk of these classes myself!

data trace command, scripts, PERL, makefiles, hardware emulation are some of the topics discussed on the list. It is also a forum for debating the content of this book as well as future books, papers and articles on verification. This list is to verification what John Cooley's *esnug*[2] is to synthesis.

Tell me where I erred.

Despite the best effort of several reviewers and many read-throughs, there are errors in this book. From simple grammatical errors in the text, to syntax errors in the code samples, to functional bugs in the algorithms. I maintain a list of errors that were found in this edition of the book in the *errata* section at:

`http://janick.bergeron.com/wtb`

If you find an error that is not listed, please let me know via email. They will be corrected in the next edition.

---

2. John can be reached at `jcooley@world.std.com`.

INDEX

A

Abstraction
  granularity for verification 70
Abstraction of data 100–125
  files 121
  lists 115
  multi-dimensional arrays 112
  using records for packets or frames 105
  verifying DSP designs 101
Arrays 112
  faking in Verilog 113
  generic implementation in Verilog 114
  multi-dimensional 112
ASIC verification 67
Assigning values 137
Automation
  eliminating human error 6
  when to use 3

B

Behavioral HDLs 83–153
  behavioral model benefits 85
  behavioral-thinking example 85
  code structure 92–99
  costs for optimizing 88
  data abstraction 100–125
  improving maintainability 91
  parallelism 125–140
  portability of Verilog 140–153

race conditions
  avoiding 147
  initialization 146
  read/write 141
  write/write 144
  RTL-thinking example 85
Behavioral models 269–289
  benefits of 85, 286
  characteristics of 273
  compared to RTL models 270
  cost of 286
  encapsulating
    bus-functional models 97
    subprograms 94
    technique 93
  equivalence with RTL models 289
  example of 271
  modeling reset in 276
  speed of 285
  writing good models 281
Black-box verification 12
Board-level verification 68
Bus-functional models
  client/server processes 241
    abstracting procotol 243
  managing client/server control
    signals 246
  multiple server instances 247
  packaging for reuse
    in Verilog 228

in VHDL 238
packaging in Verilog 99
packaging in VHDL 98
using qualified names 246

C

Capitalization
naming guidelines 330
Code coverage 40–46
and testbench generators 11, 46
code-related metrics 57
expression coverage 45
path coverage 44
statement coverage 42
usefulness 45
Code reviews 28
Coding guidelines 317–345
Comments
guidelines 321
quality of 91
Component-level features 73
Concurrency
definition of 125
misuse of 132
problems with 127
with fork/join statement 134
Connectivity
definition of 125
Constants
naming guidelines 334
Co-simulators 34
Costs for optimizing 88
Cycle-based simulation 31

D

Data abstraction 100–125
arrays 112
faking records in Verilog 107
files 121
lists 115
real values 101
records 105
Delta cycles 130
Driving values 138

E

Encapsulating
bus-functional models 97
subprograms 94
technique 93
Equivalence checking 8
Error types 74
Event-driven simulation 29
Expression coverage 45

F

False negative 18
False positive 18
File-driven testbenches 259
Filenames
guidelines 336
Files 121
external configuration 260
external input 259
name uniqueness 261
fork/join statement 134
Formal verification 7
equivalence checking 8
model checking 9
FPGA verification 67
Functional verification
black-box 12
grey-box 13
purpose of 10
white-box 13

G

Generics
top-level 266
Grey-box verification 13
Guidelines
capitalization 330
code layout 323
code syntax 326
comments 321
constants 334
debugging 329
directory structure 318–321
filenames 336
general coding 321–329
HDL code layout 337
HDL code structure 337

HDL coding 336–345
HDL specific naming 334
identifiers 332
naming 329–336
Verilog specific 340
VHDL specific 337

H

Hardware modelers 37

I

Identifiers
naming guidelines 332
Issue tracking 52–57
computerized system 55
grapevine system 53
Post-it system 54
procedural system 55

L

Linked lists 117
Linting tools 22–28
code reviews 28
limitations of 23
with Verilog 25
with VHDL 26
Lists 115
in Verilog 119

M

Maintaining code
commenting 91
optimizing 88
Metrics 57–60
code-related metrics 57
interpreting 59
quality-related metrics 58
Model checking 9

N

Naming
capitalization guidelines 330
constants 334
filenames 336
guidelines 329–336
HDL specific guidelines 334

identifiers 332

O

Output, predicting 211–219
complex transformations 216
data formatters 211
packet processors 215
using external C program 259

P

Packaging bus-functional models
in Verilog 99
in VHDL 98
Parallelism 125–140
concurrency problems 127
driving vs assigning 137
emulating 128
implementation differences 126
misuse of concurrency 132
simulation cycle 129
Parameters
setting top-level parameters
in Verilog 267
in VHDL 267
top-level 266
Path coverage 44
Poka-yoka 6
Portability of Verilog
non-reentrant tasks 151
race conditions
avoiding 147
initialization 146
read/write 141
write/write 144
using disable statement 148
using ouput arguments 150
Procedural interface
verification components 225
Profiling 46
Programmable testbenches 259

R

Race conditions
avoiding 147
initialization 146
read/write 141

read/write and synchronized
    waveforms 161
write/write 144
Random verification 71
Real numbers
    fixed-point representation 104
    limitations in Verilog 102
Reconvergence model 4–5
Records 105
    faking in Verilog 107
    variant records 111
Redundancy 6, 81
Regression 185
    management 314
    running 313
Regression testing
    for reusable components 67
Resolution functions 136
Response
    autonomous error detection 258
    autonomous monitoring 255
    verifying 70
Response, complex 193–210
    abstracting ouput operations 199
    definition of 194
    generic output monitors 202
    handling latency 195
    monitoring bi-directional
        interfaces 205
    monitoring multiple operations 203
Response, verifying 172–176
    inspecting response visually 172
    inspecting waveforms visually 174
    minimizing sampling 174
    sampling output 172
Reuse
    and verification 16–17
    level of verification 64
    packaging bus-functional models
        in Verilog 228
        in VHDL 238
    reusable verification components 221–
        227
    trust 16
    utility packages
        in Verilog 231
        in VHDL 249
    verification of components 66

Revision control 47–52
    configuration management 50
    working with releases 51

S

Scan-based testing 14
SDF back-annotation 305
Simulation cycle 129
Simulation management 269–316
    configuration management 292–312
    configuration management in
        verilog 295
    configuration management in
        VHDL 301
    determining success or failure 289–292
    output files 309
    regression 312–316
    SDF back-annotation 305
Simulators 28–35
    co-simulators 34
    cycle-based simulation 31
    event-driven simulation 29
    single-kernel 35
    stimulus and response 29
Sparse memory model 117
    damem PLI package 119
    using a linked list in VHDL 117
Statement coverage 42
Stimulus
    autonomous 250
    injecting errors 255
    random 253
Stimulus, complex 183–193
    asynchronous interfaces 187
    configurable operations 192
    CPU operations 189
    deadlocks 184
    feedback between stimulus and
        design 183
Stimulus, simple 155–171
    abstracting waveform generation 169
    aligning waveforms 164
    encapsulating waveforms 167
    generating
        complex waveforms 159
        simple waveforms 156
        synchronized waveforms 160
        synchronous data waveforms 165

Structure
    Coding guidelines for 337
System-level features 73
System-level verification 68

T

Test harness 224
    in Verilog 228
    in VHDL 240
    multiple bus-functional server
        instances 247
Testbench generators
    and code coverage 11, 46
Testbenches
    and formal verification 7
    configurable 265
    generators 11
    grouping by testcases 79–81
    verifying 80
Testbenches, architecting 221–268
    abstracting client/server protocol 243
    autonomous error detection 258
    autonomous generation and
        monitoring 250–258
    autonomous monitoring 255
    autonomous stimulus 250
    compile-time configuration 262
    concurrent simulations 261
    configurable testbenches 265
    configuration files 260
    file-driven testbenches 259
    in Verilog 227–237
    in VHDL 237–249
    injecting errors 255
    inputs and outputs 258–263
    managing client/server control
        signals 246
    multiple server instances 247
    packaging bus-functional models
        in Verilog 228
        in VHDL 238
    programmable testbenches 259
    random stimulus 253
    test harness in VHDL 240
    utility packages in Verilog 231
    utility packages in VHDL 249
    verifying configurable designs 263–
        268

Testbenches, self-checking 176–182
    golden vectors 177
    input and output vectors 176
    run-time verification 179
Testcases
    grouping by code coverage metrics 76
    grouping by feature 76
    grouping into testbenches 79–81
Testing
    and verification 13–16
    scan-based 14
Third-party models 36–38
    hardware modelers 37
Time
    definition of 125
Top-level generics 266
Top-level parameters 266
Type I error 18
Type II error 18

U

Unit-level verification 65

V

Verification
    and design reuse 16–17
    and testing 13–16
    ASIC and FPGA 67
    black-box verification 12
    board-level 68
    checking result of transformation 4
    components 221–227
    cost 17
    definition of 1
    designing for 16
    formal verification 7
    functional verification 10, 11–13
    grey box verification 13
    importance of 2–4
    improving accuracy of 6, 81
    need for specifying 62
    plan 61–81
    purpose of 4
    random strategy for 71
    reducing human error 5
    reusable components 66, 221–227
    strategies for 69–72
    system-level verification 68

testbenches, verifying 80
tools 21–60
types of mistakes 17
unit-level verification 65
white-box verification 13
with reconvergence model 4–5
Verification components
development process 226–227
in Verilog 227–237
in VHDL 237–249
procedural interface 225
Verification languages 46–47
Verification plan
component-level features 73
definition of 63
design for verification 77
error types 74
grouping features into testcases 76
grouping testcases into testbenches 79–81
identifying features 72–75
levels of verification 64–69
prioritizing features 75
role of 62–81
strategies for verification 69–72
system-level features 73
verifying testbenches 80
Verification strategies 69–72
random verification 71
verifying the response 70
Verification tools 21–60
code coverage 40–46
issue tracking 52–57
linting 22–28
metrics 57–60
revision control 47–52
simulators 28–35
third-party models 36–38
verification languages 46–47
waveform comparators 39
waveform viewers 38–40
Verilog
coding guidelines 340
configuration management
guidelines 320
recommended textbook xix
vs VHDL xx–xxi
VHDL
coding guidelines 337

recommended textbook xix
vs Verilog xx–xxi

W

Waveform comparators 39
Waveform viewers 38–40
limitations of 39
Waveforms
abstracting generation 169
aligning 164
encapsulating 167
generating
complex waveforms 159
simple 156
synchronized waveforms 160
synchronous data waveforms 165
White-box verification 13